System Design with SystemC™

System Design with SystemC™

Thorsten Grötker Stan Liao
Synopsys, Inc.

Grant Martin Stuart Swan
Cadence Design Systems, Inc.

Kluwer Academic Publishers
Boston / Dordrecht / London

Distributors for North, Central and South America:
Kluwer Academic Publishers
101 Philip Drive
Assinippi Park
Norwell, Massachusetts 02061 USA
Telephone (781) 871-6600
Fax (781) 681-9045
E-Mail: kluwer@wkap.com

Distributors for all other countries:
Kluwer Academic Publishers Group
Post Office Box 322
3300 AH Dordrecht, THE NETHERLANDS
Telephone 31 786 576 000
Fax 31 786 576 254
E-Mail: services@wkap.nl

Electronic Services < http://www.wkap.nl>

Library of Congress Cataloging-in-Publication Data
System design with System C/Thorsten Grötker....[et al].
p.cm.
Includes bibliographical references and index.
ISBN 978-1-4419-5285-1 e-ISBN 978-0-306-47652-5
1.System design. 2. C++ (Computer program language) I.Grötker, Thorsten
QA76.9.S88 /S9525 2002
004.2'1--dc21 2002021534

Softcover reprint of the hardcover 1st edition 2002

Printed on acid-free paper.

This printing is a digital duplication of the original edition.

Contents

Figures

Foreword

I am honored and delighted to write the foreword to this very first book about SystemC. It is now an excellent time to summarize what SystemC really is and what it can be used for.

The main message in the area of design in the 2001 International Technology Roadmap for Semiconductors (ITRS) is that "cost of design is the greatest threat to the continuation of the semiconductor roadmap." This recent revision of the ITRS describes the major productivity improvements of the last few years as "small block reuse," "large block reuse," and "IC implementation tools." In order to continue to reduce design cost, the required future solutions will be "intelligent test benches" and "embedded system-level methodology."

As the new system-level specification and design language, SystemC directly contributes to these two solutions. These will have the biggest impact on future design technology and will reduce system implementation cost. It took SystemC less than two years to emerge as the leader among the many new and well-discussed system-level design languages. In my opinion, this is due to the fact that SystemC adopted object-oriented system-level design—the most promising method already applied by the majority of firms during the last couple of years.

Even before the introduction of SystemC, many system designers have attempted to develop executable specifications in C++. These executable functional specifications are then refined to the well-known transaction level, to model the communication of system-level processes. Transactions are nonatomic communications, normally with bidirectional data transfer, and consist of a set of messages that are usually modeled as atomic communications. The messages have unidirectional data transfer, but often bidirectional control flow, and are then implemented at a cycle-true and bit-accurate signal level.

SystemC has successfully consolidated the design flow in use at many firms for some time, and standardized it by making available corresponding libraries. In this respect, SystemC follows the principle of defining a language not only by its syntax and semantics, but also by specification of various libraries. This principle was very successfully introduced by the Java language.

Despite the fact that SystemC is a system-level design and specification language, the book gives a detailed introduction to traditional hardware modeling. It is important, in my experience, to stress cycle-true hardware modeling in SystemC, since this facilitates the real adoption of SystemC by VHDL and Verilog designers. Equally important, the book also makes it clear how SystemC designs and testbenches can be interfaced with existing HDL designs.

SystemC takes up the challenge of being compared to the traditional hardware description languages. For the future success of SystemC, it is essential that it can be used reliably on all levels: system level, register-transfer level, and those in between. Numerous examples in the book give proof that SystemC can be used successfully in these ways. The capability of SystemC for modeling at register-transfer level and behavioral level is demonstrated. Functional modeling with dataflow is introduced. The discussion on the design of interfaces and channels, which lay the foundation for higher levels of abstraction, leads on to transaction-based design.

The book describes both this bottom-up abstraction, and its opposite: top-down refinement for design processes starting at the system level. For this purpose, I find the chapter on "communication refinement" very useful. As the organizer and moderator of the European SystemC User Group, whose meetings are a good forum for many designers using SystemC, I know that this topic provokes frequent discussion. This chapter gives many useful explanations of communication refinement which go well beyond the SystemC manuals. Communication refinement serves as a good framework for guidelines on "practical design in SystemC," which the book introduces clearly.

In addition to hardware–hardware refinement, the book also describes software–software and hardware–software refinement, topics of considerable current interest. In the general design community, these subjects are frequently discussed, but with few facts. This book goes beyond general discussion and provides specific and useful information in this area.

The same issue holds for topics such as intellectual property (IP), and especially IP reuse. Here, the chapter on parameterizable modules and channels shows that SystemC is no longer in its experimental phase but is a real language with real tools and allows real IP reuse. The discussion on

IP protection stresses professional and commercial aspects of SystemC and their growing importance.

Finally, the book does not miss important topics; it gives both an introduction to models of computation, and, on the other hand, very practical hints with respect to testing and debugging, and avoiding traps and pitfalls in the use of SystemC.

In summary, this is currently the only book providing a complete description of what SystemC really is, and of what can be actually achieved by putting it to use. The book serves both as an introduction to SystemC for students and practitioners as well as a valuable companion for engineers dealing with the design and verification of digital embedded systems at various levels of abstraction. Therefore, I would urge everyone interested in SystemC to read this book, and apply its lessons.

Wolfgang Rosenstiel
Tübingen, Germany
March 2002

Acknowledgments

SystemC would not be the best possible language today for system-level design and exploration if it were not for the collective efforts of many people over a number of years. This book is therefore not only the fruit of the authors' labors, but also a reflection of the efforts of the members of the Language Working Group (LWG) of the Open SystemC Initiative (OSCI). We would like to thank the LWG members for their hard work and dedication.

However, there are some people who have helped in particular with this book and with SystemC whom we would like to acknowledge by name. These include Abhijit Ghosh for his technical leadership and inspiration; Martin Janssen for helping to define SystemC 2.0 and his key role in creating a high quality SystemC 2.0 reference simulator; Ric Hilderink for developing the `simple_bus` transaction-level model that is discussed within chapter 8; and Kevin Kranen and Stan Krolikoski for recognizing the great need for a standard system level modeling language and for their hard work in making SystemC a reality through the formation of OSCI and the enlistment of member companies.

A special acknowledgment goes to Dan Gajski, whose work on SpecC methodology for many years was an important reference point for SystemC as it developed system-level modeling abstractions.

We would also like to acknowledge Wolfgang Rosenstiel for writing the foreword, and the following people for their review of the book and valuable feedback: Abhijit Ghosh, Joe Buck, Norris Ip, Martin Janssen, Holger Keding, Rajesh Kumar, Simon North, and Bob Shur.

We thank our employers, Synopsys, Inc., and Cadence Design Systems, Inc., for their support of SystemC and the technical work that has been involved in bringing it to its current state.

Finally, we would like to thank Stephen Edwards of Columbia University for his LaTeX template (from his book *Languages for Digital Embedded Systems*), which was an invaluable starting point for our book production; and Dan Luecking of the University of Arkansas for his tip on a LaTeX technique. We are grateful to Michael Rauner (*m@michaelrphotography.com*) for the cover photography.

Thorsten Grötker
Stan Liao
Grant Martin
Stuart Swan
March 2002

* * *

Thorsten Grötker dedicates this book to the SystemC community and his colleagues and family who enabled him to work on the book.

Stan Liao dedicates this book to his parents, Mei-Sung and Chih-Mei Liao.

Grant Martin dedicates this book to his wife, Margaret Steele, and his daughters, Jennifer and Fiona Martin.

Stuart Swan dedicates this book to his wife Mary and his children, Thomas and Elizabeth.

1

Introduction

1.1 Motivation

What is the motivation for introducing a new modeling language such as SystemC? Why is it insufficient for designers to keep using familiar hardware description languages (HDLs), such as Verilog and VHDL, which are incorporated into well-proven and well-understood design flows?

Our motivation starts with the changing nature of "systems" under design. When systems were composed primarily of discrete parts such as microprocessors, memory chips, analog devices, and application-specific integrated circuits (ASICs), the design process usually started with one or two system design experts who would partition the functionality into hardware and software, and further partition the hardware function into standard parts and ASICs. A specification for an ASIC of a few thousand to a few hundred thousand gates was possible to write in a natural language, and hand off to an ASIC designer or team, who would start by capturing the design at the register-transfer level (RTL) for which HDLs were the perfect match.

Contrast this with the modern system-on-chip (SoC) [7]. Such an integrated device may well contain one or more processors including both 32-bit microcontrollers and digital signal processors (DSPs) or specialized media processors. On-chip memory, accelerating hardware units for dedicated functions, and peripheral control devices, linked together by a complex on-chip communications network that incorporates on-chip busses, complete the definition. Software and its architecture, layering, and complexity are inherent in such a design. Clearly, the definition of a "system" has moved on from the system-on-board days.

To specify, design, and implement such complex systems, incorporating functionality implemented in both hardware and software forms, we are compelled to move on from the HDLs of old. We must also move beyond the RT level of abstraction used with these HDLs. We need to move to what has been termed the "system-level" of design. And we need a modeling language that can support this level.

What are the requirements of such a language? It must support:

- specification and design at various levels of abstraction;
- incorporation of embedded software portions of a complex system, both models and implementation-level code;
- creation of executable specifications of design intent;
- creation of executable platform models, representing possible implementation architectures on which the design intent will be mapped;
- fast simulation speed to enable design-space exploration, of both functional specification and architectural implementation alternatives; and
- constructs allowing the separation of system function from system communications, in order to allow flexible adaptation and reuse of both models and implementations.

In addition, such a language should ideally be based on a well-established programming language, in order to capitalize on the extensive infrastructure of capture, compilation, and debugging tools already available. An object-oriented language base would allow for modeling flexibility and facilitate reuse, through capabilities such as templates and inheritance. Although arguably not the easiest language to learn, C++ has proven to be a reasonable choice as the basis for such a system-design language. Java might have been an interesting alternative, but was less mature than C++, had few trial implementations of system level design concepts to work with, was used by many fewer people in the system design area, and, in its interpreted form, would have had serious performance issues in comparison to C++.

The absence of a standardized and well-accepted system-level design language has inhibited the development, exchange, and reuse of intellectual property (IP) models at the system level. Where reuse occurs in hardware design, it starts at the RT level using standard HDLs. Because IP modeled in HDLs is usually done with implementation in mind, this inhibits design space exploration and is relatively poor in IP protection, thus preventing easy evaluation. RT-level models also simulate slowly compared to more abstract models. The Virtual Socket Interface Alliance (VSIA) System-Level Design Development Working Group has defined the semantics of a system-level interface format (SLIF) which would facilitate IP exchange at the system level [1] [21]. SLIF defines principles for system-level design languages to support a methodology in which functionality and interface definitions are separated, and interfaces can be continuously refined from a high level of abstraction to a very detailed implementation level. It uses concepts including abstract and concrete data-types and ports, interface attributes, and a transaction/messaging concept—which all map very well into SystemC.

To summarize, the fundamental motivator for SystemC is to provide a modeling framework for systems in which high-level functional models can be refined down to implementation in a single language. This is a major contribution to the development of system-level design.

1.2 How to Read This Book

This book is an introduction to SystemC and its use for system-level design, modeling, and refinement. The book is based on SystemC version 2.0 as defined by the documents [36] [37].

We assume the reader is familiar with C and C++. References such as Meyers [24] and Stroustrup [33] should be consulted for details. We also assume that readers interested in hardware modeling are familiar with hardware description languages (HDLs) such as Verilog and VHDL; please consult references such as Moorby and Thomas [27] for more details on Verilog, and Lipsett et al. [23] for more details on VHDL. Of course, readers focusing on high-level modeling can concentrate on the system-level aspects of SystemC, and skip lightly past the areas dealing with HDL-type constructs. A few sections in this book that involve more complicated material are denoted by an asterisk (*) before the section headings.

Although SystemC is intended as a medium for system-level design, this book is primarily about the language and its usage, not a comprehensive book about system-level design concepts and methodologies. We recommend sources such as De Micheli et al. [26] for information about hardware–software co-design; Balarin et al. [2] on co-design of embedded systems; and Gajski et al. [12] for related methodologies and languages for system-level design.

We deal with the core SystemC 2.0 language and, in particular, are not covering the methodology-specific libraries based on the core language features. However, we will discuss how the core language features can be used to build methodology-specific libraries and the use-methodologies for them.

1.3 History

SystemC is the confluence of four streams of ideas: work at Synopsys with University of California, Irvine, and later with Infineon (formerly Siemens HL) as well; Frontier Design; IMEC; and work within the Open SystemC Initiative (OSCI) Language Working Group [34] [35].

The SCENIC project at Synopsys/UC Irvine [22] built a design environment in C++ for modeling hardware and software components of systems, and emphasized the modeling of hardware components using classes to create hardware-oriented data-types. SCENIC also provided capabilities for modeling and simulating hardware using a reactive, cycle-based approach. As SCENIC evolved, it incorporated certain simulation and tracing techniques from Infineon.

From Frontier Design (now Adelante Technologies, after a merger with the DSP division of Philips Semiconductor) came contributions on data-types, especially fixed-point data-types derived from their work on signal processing systems. This work was done in common with the VSIA SLD DWG work on system-level data-types [1] [21].

From IMEC came work on hardware–software co-design (now marketed by CoWare) [29], and C/C++-based design that had an influence on early SystemC development.

Most importantly, the Language Working Group of OSCI has developed the main system-level design concepts in SystemC, especially from mid-2000 to mid-2001 for the 2.0 specification. These include an improved model of time, dynamic sensitivity, user-defined interfaces, channels and ports, with primitive and hierarchical channels, as well as language restructuring separating the core of the language from methodology-specific libraries.

SystemC is not alone in advocating C/C++ system-level design concepts. SpecC [12] shares similar concepts of abstract functional and communications modeling of systems with staged refinement towards implementation, and indeed was a useful reference for the OSCI Language Working Group. IMEC has continued its work on C++-based design with OCAPI

[32]. Another C++ library approach is Cynlib, offered by CynApps (now Forte Design Systems) [28]. However, recently (as of late 2001) there has been an important convergence of the industry onto SystemC as the leading approach to system-level modeling with C++. Forte has joined OSCI and will be contributing ideas based on their Cynlib experiences. SystemC 2.0 was influenced by concepts in SpecC, and the "father" of SpecC, Professor Daniel Gajski of University of California, Irvine, is involved in useful discussions with the OSCI Language Working Group. There has been growing interest in Europe in SystemC, with a very active European SystemC Users Group led by Professor Wolfgang Rosenstiel of the University of Tübingen [11].

1.4 Modeling Levels

The focus of this book is on the SystemC language and how to apply the SystemC language in a wide variety of modeling tasks. The goal of this book is not to advocate specific design methodologies, nor is it to advocate specific modeling levels or terminology. Nevertheless, for the discussion in subsequent chapters it is necessary to introduce several important concepts and terms related to different modeling levels.

First, it is important to recognize that SystemC does not impose a top-down or bottom-up or even middle-out design flow. In fact it is recognized that most design flows are iterative, and that it is rare that all modules within a system are modeled at the same level of abstraction. We commonly hear from designers in industry that real designs hardly ever start with a "clean sheet of paper," so the need to model testbenches and preexisting hardware and software implementations at various levels of abstraction while adding new models, potentially at different levels of abstraction, is quite common.

Let's list a few simple design scenarios where different modeling levels might be used:

- A designer might use a very detailed implementation-level model for a design under test while using abstract models within the testbench to generate the design's stimulus and check the response.
- With a detailed implementation-level model as a starting point, a designer might create a more abstract model in order to increase simulation speed and perhaps to protect intellectual property that might otherwise be exposed within the more detailed model.

- A designer might refine a module from a high-level functional specification down to a cycle-accurate RTL model while other modules in the system remain at higher levels of abstraction.

When we consider a particular SystemC model and compare it to an existing or proposed real-world implementation, we note that there are several independent axes on which we can gauge the model's accuracy. These include:

- *structural accuracy:* The extent to which the model reflects the structure of the actual implementation. For example, whether the partitioning between hardware and software is modeled and whether major hardware and software modules within the implementation are evident; for hardware models whether the signals and pins of the actual implementation are reflected in the model; and for software models whether communication between tasks is refined down to the level of communication mechanisms provided by the target real-time operating system (RTOS).

- *timing accuracy:* The extent to which the model reflects the timing of the actual implementation. Timing effects may arise because of constraints imposed by the design specification, and they will also arise because of processing delays within the target implementation. Timing may be expressed in absolute time units (e.g., seconds) with approximate or exact values. Often with hardware models we only care about timing at the clock-cycle level and are not concerned about timing below that level, since hardware implementation tools help manage such details for us.

- *functional accuracy:* The extent to which the model reflects the function of the actual implementation. In some cases a high-level model may omit certain complex functionality to simplify the model or increase simulation speed, despite the fact that this functionality is present in the implementation. As an example, in digital signal processing designs it is common to create initial functional models that use floating-point data-types and arithmetic. After initial functionality has been verified, this functional model is refined to a "bit-true" or "bit-accurate" model that accurately models the effects of fixed-precision data-types and arithmetic. (See section 6.3.1 for an example of fixed-point modeling in SystemC.)

- *data organization accuracy:* The extent to which the model reflects the actual data organization used within the implementation. For example, whether the software data structures and the data layout within the model match those used within the implementation.
- *communication protocol accuracy:* The extent to which the model reflects the actual communication protocols used within the target implementation. In some modeling scenarios it is useful to replace detailed communication protocol models with more abstract and efficient high-level models.

For each of the different modeling accuracy aspects above, sometimes we also need to distinguish whether we are talking about the particular accuracy aspect only at a module's boundaries (i.e., at the module's ports), or whether the accuracy aspect also extends to all child modules contained within the parent module.

We also note that the modeling aspects listed above apply to software as well as hardware models. With software models we also see that it is important to identify the model accuracy in terms of structure, timing, function, data organization, and communication protocols.

Now that we have identified some of the important aspects which determine model accuracy, we can introduce some of the terms that we use within this book to describe different modeling levels.

An *executable specification* is a model that is a direct translation of a design specification into SystemC. Executable specifications model the intended functionality of a design in a manner that is completely independent of any proposed implementation. If time delays are present in an executable specification, they represent timing constraints to be imposed on the implementation.

An *untimed functional model* is similar to an executable specification, but no time delays at all are present in the model. Communication between modules within an untimed functional model is point-to-point (i.e., no shared communication links such as busses are modeled). Usually the communication is modeled using FIFOs with blocking write and read methods so that data items are reliably delivered between modules.

A *timed functional model* is similar to an untimed functional model in that communication between modules is still point-to-point (i.e., still no modeling of shared communication links) and in that it typically uses FIFOs with blocking read and write methods. However, in a timed functional model timing delays are added to processes within the design to reflect the timing constraints of the specification and processing delays of a particular target implementation. In addition, timing delays may be annotated onto FIFOs used for communication between modules in order to model communication delays of a target implementation. Timed functional models are often used for early hardware–software trade-off analysis. This is done by evaluating the impact of mapping processes to hardware (resulting in one delay figure) versus to software (resulting in another delay figure). Timed functional models can also be used to determine effects that latency may have on control-oriented algorithms.

Note that executable specifications and both untimed and timed functional models do not have any direct structural correspondence to a target implementation. We will describe how executable specifications and both untimed and timed functional models can be created in SystemC in chapter 5.

In a *transaction-level model* communication between modules is modeled using function calls. In such models, the communication is typically modeled in a way that it is accurate in terms of functionality and often in terms of timing (sometimes even accurate to the clock-cycle level), but the communication is *not* modeled in a way that is structurally accurate. For example, in a transaction-level model of a system-on-chip (SoC) platform, we might model the different types of transactions that the on-chip bus supports (e.g., burst read/write transactions), but we don't model the actual wires of the bus or the pins of the modules that connect to the bus.

When we use the term *platform transaction-level model* we are indicating that a model uses both the transaction-level modeling style described above to model the communications infrastructure of a target SoC platform, and that the modules within such a design structurally correspond to blocks within the target implementation. Platform transaction-level models can be used to accurately model effects such as bus loading and contention and overall system performance, and they also provide a high-performance, accurate and effective way to model hardware and software interactions at a very early stage in the design process. We will discuss transaction-level models at length in chapter 8 and chapter 9.

A *behavioral hardware model* is a model that is pin-accurate and functionally accurate at its boundaries, but which is not considered to be clock-cycle accurate at its boundaries. A behavioral hardware model does not have internal structure that reflects the structure of the target implementation. Behavioral hardware models can be used as input to hardware behavioral synthesis tools, and will be discussed in chapter 4.

A *pin-accurate, cycle-accurate hardware model* is a model that has these two characteristics at its boundaries, in addition to being functionally accurate. Such a model does not necessarily have internal structure that reflects the target implementation.

Like a pin-accurate, cycle-accurate hardware model, a *register-transfer (RT) level model* has these characteristics at its boundaries. In addition, the internal structure of an RT level model accurately reflects the registers and combinational logic of a target implementation. We will discuss RT level modeling in SystemC in chapter 4.

1.5 Summary

In this chapter we have briefly introduced the background of SystemC to the reader. We have covered the prerequisites assumed for comprehension and use of this book. The motivation for SystemC development and a brief survey of its history have been described. A summary description of some key modeling abstraction levels has been given; these will be further elaborated later in this book. We are now ready to consider the language features and capabilities in more detail.

1.6 SystemC Resources

Readers who wish to find out more about SystemC beyond what is covered in this book are advised to use the following resources:

- The Open SystemC Initiative (OSCI) web site at URL:
  http://www.systemc.org
- The European SystemC Users Group web site at URL:
  http://www-ti.informatik.uni-tuebingen.de/~systemc
- The SystemC discussion forum mailing list at the OSCI site:
  http://www.systemc.org

We also recommend that the reader visit the web page for this book, located in the OSCI web site *http://www.systemc.org*. Look in the section *Products & Solutions* ▷ *SystemC Books* for the link to this page (*System Design with SystemC*). There you will find source code downloads for all of the examples in this book, any clarifications and errata, and information about updated editions. In addition, the authors of this book would be happy to hear feedback from readers on its content and suggestions for improvement in future editions; you may contact the authors also via this web page.

2

Fundamentals of SystemC

2.1 Introduction

One of the primary goals of SystemC is to enable system-level modeling—that is, modeling of systems above the register-transfer levelof abstraction, including systems that might be implemented in software, hardware, or some combination of the two. Of course RTL modeling can also be performed in SystemC, and, in addition, one can develop models above the RT level and refine them down to RTL within a single language and environment.

The wide range of models of computation and communication, levels of abstraction, and methodologies used in system design presents a major challenge in creating a system-level design language. For instance, certain DSP problems are naturally mapped to a dataflow model of computation (of which Kahn process networks are a generalization; see chapter 3); and many reactive control systems are commonly modeled with finite automata. Furthermore, a large system may be composed of communicating subsystems that are modeled heterogeneously, and such a system needs to be simulated efficiently.

To address this challenge, SystemC uses a layered approach that allows for the flexibility of introducing new, higher-level constructs that share an efficient simulation engine. The base layer of SystemC provides an event-driven simulation kernel. This kernel works with events and processes in an abstract manner. It knows only how to operate on events and switch between processes, without knowing what the events actually represent or what the processes do. Other elements of SystemC include modules and ports for representing structural information, and interfaces and channels

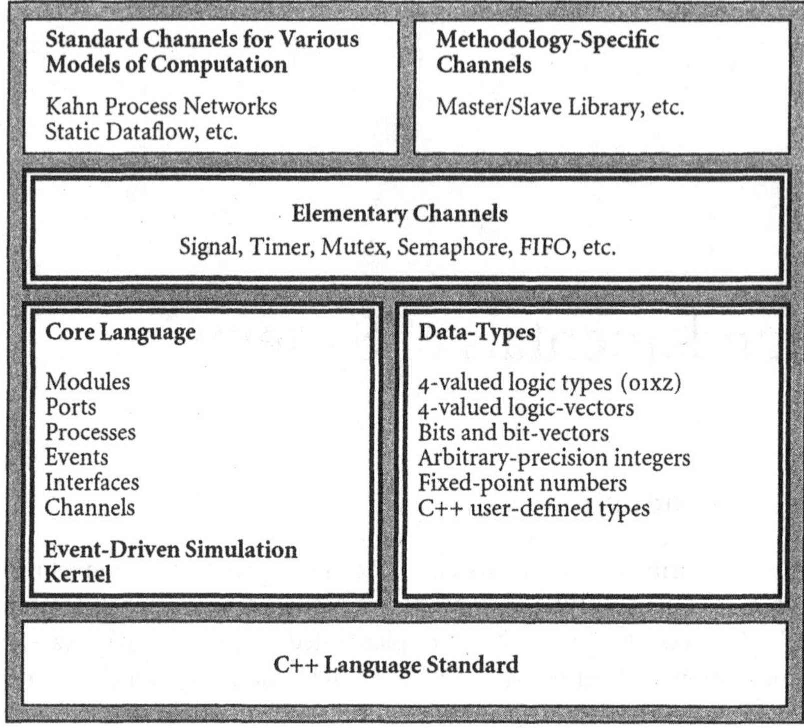

Figure 2.1: SystemC language architecture.

as an abstraction for communication. The kernel and these abstract elements together form the *core language*. On top of this language foundation we can then add the more specific models of computation, design libraries, modeling guidelines, and design methodologies that are useful for system design. Note that the core language itself, along with elementary channels and SystemC data-types, is sufficiently powerful to model many systems.

Figure 2.1 shows the various layers of SystemC. The bottommost layer highlights the fact that SystemC is built entirely on standard C++. This means that any program written in SystemC may be compiled with a C++ compiler to produce an executable program. Alongside the core language is a set of data-types that are useful for hardware modeling and for certain kinds of software programming (e.g., fixed-point types for DSP software, and for hardware implementations of DSP-type functions). In chapter 4 we will describe the SystemC data-types in more detail.

It should be noted that not all of the blocks shown in figure 2.1 are considered part of the SystemC standard. The small, general-purpose core language lays the foundation for SystemC and is the central component of the SystemC standard. The elementary channels, depicted in the layer immediately above the core language, include models that are widely applicable, such as signals, timers, and first-in-first-out buffers (FIFOs). For this reason, we treat them as components of the SystemC standard as well, though the distinction of layers is maintained. Likewise, although the hardware data-types are orthogonal to the core language, we include them in the standard because of their broad usefulness. These parts are emphasized with double borders in figure 2.1.

On the other hand, we recognize that many different models of computation and design methodologies may be used in conjunction with SystemC. Hence, the design libraries and models needed to support these specific design methodologies are considered to be separate from the SystemC standard. For instance, the channels shown in the topmost layers are not part of the standard; but the designer may choose, as he sees fit, to use them, or even devise novel components for new models of computation. Over time, as new methodologies prove to be useful and effective, they may themselves become standardized and may be incorporated into SystemC, either into the core language or as libraries that layer on top of the core language.

In the remainder of this chapter, we will describe the elements of the SystemC core language. We will examine the function and usage of each element, and finally, the semantics of the event-driven simulation kernel. In this chapter and throughout the book, we will use C++ terminology to explain SystemC constructs. In addition, we will use terminology from traditional hardware modeling and simulation for concepts that have counterparts in that domain. For instance, we use the term *elaboration* to refer to that phase of execution in which the SystemC library routines undertake the preparatory work to construct and connect the objects for simulation, as prescribed by the designer.

2.2 Model of Time in SystemC

SystemC uses an integer-valued time model. In the current reference implementation, the underlying data-type, as in VHDL and Verilog, is a 64-bit unsigned integer; the size can be increased to more than 64-bit if necessary. Since simulation time is discrete, there is a minimum resolvable time unit (*time resolution*), the smallest quantum of time that can be represented. Any time value smaller than the time resolution will be rounded off. The default value of time resolution is one picosecond (10^{-12}s), and the user has the option of setting it to some other value by calling `sc_set_time_resolution()`. The time resolution must be a power-of-ten multiple of a second, and must be specified before any time objects (of class `sc_time`) are created, normally before elaboration begins. To capture the notion of time more precisely, SystemC defines an enumerated type `sc_time_unit` with the following values and attributed meanings:

SC_FS	femtosecond
SC_PS	picosecond
SC_NS	nanosecond
SC_US	microsecond
SC_MS	millisecond
SC_SEC	second

The `sc_time` type is used to represent time or time interval in SystemC. Time objects are constructed with a numeric value (of type `int` or `double`) and a time unit. For instance,

```
sc_time t1(42, SC_PS);
```

creates a time object `t1` representing 42 picoseconds. The intuitive meaning of a time object does not depend on its internal representation, which may vary because of changes in time resolution. Also, time objects are susceptible to rounding errors. For example, if we declare `t2` as follows after the time resolution has been changed to 10 ps:

```
sc_set_time_resolution(10, SC_PS);
sc_time t2(3.1416, SC_NS);
```

then the actual value of `t2` would be rounded to 3140 ps.

2.3 Modules

Modules are the basic building blocks for partitioning a design. They allow designers to break complex systems into smaller, more manageable pieces, and to hide internal data representation and algorithms from other modules. A typical module contains:

1. ports, through which the module communicates with the environment;

2. processes that describe the functionality of the module;
3. internal data and channels for maintenance of model state (e.g., state number of a finite automata) and communication among the module's processes; and
4. hierarchically, other modules.

A module is described with the SC_MODULE macro, with the body of the module enclosed in curly braces. For example:

```
SC_MODULE(Adder) {
    // ports, processes, internal data, etc.
    SC_CTOR(Adder) {
        // body of constructor:
        // process declaration, sensitivities, etc.
    }
};
```

There is nothing magical about the SC_MODULE macro; it is simply shorthand for deriving class Adder from the library class sc_module. This is the simplest and most common form of module definition. More advanced forms such as those involving user-defined inheritance require explicit derivation. For instance, the class hw_fifo_wrapper in section 7.5 is derived explicitly from sc_module, as well as two other classes. The macro SC_CTOR declares a *constructor*, which maps designated *member functions* to *processes* and declares *event sensitivities* (section 2.7). Its argument must be the name of the module that is currently being declared, or else the compiler will complain. SC_CTOR provides a convenient way to manage the module's hierarchical names automatically, but only for the simplest form of modules. It also declares a special symbol for use with the more complex macros, SC_METHOD and SC_THREAD (section 2.5). Again, the user must spell out the declaration of the constructor if SC_CTOR does not meet his needs, or use the macro SC_HAS_PROCESS to prepare for SC_METHOD and SC_THREAD, e.g.,

```
SC_MODULE(Adder) {
    // ports, processes, internal data, etc.
    SC_HAS_PROCESS(Adder);
    Adder(sc_module_name name, int other_param) : sc_module(name) {
        // body of constructor:
        // process declaration, sensitivities, etc.
    }
};
```

Modules may be *instantiated*: there may well be two or more instances whose structure and functionality are identical, described by one module definition, e.g., two instances of the Wallace-tree multiplier. As we represent modules by C++ classes, module instantiation follows the rules for object instantiation. There are additional rules, however. For constructors declared with SC_CTOR, during instantiation the user is required to supply a name to the instance. For example, to declare an instance of Adder named xyz, we state:

```
Adder xyz("xyz");
```

Also, it is recommended that port binding (section 2.4.2) immediately follow module instantiation. In section 2.9 we will discuss module instantiation and composition.

2.4 Interfaces, Ports, and Channels

In classical hardware modeling, the hardware signal is the medium for communication and synchronization between processes. From a system-level point of view, this level of abstraction is too low. SystemC uses interfaces, ports, and channels to provide for the flexibility and high level of abstraction needed. In this scheme, channels are the workhorses for holding and transmitting data, and an interface is a "window" into a channel that describes the set of operations, or a subset thereof, that the channel provides. Ports are proxy objects that facilitate access to channels through interfaces. At elaboration time, when the SystemC library connects each port to a designated channel, we say that the port is *bound* to the channel *through* an interface.

As figure 2.1 shows, SystemC includes several "packages" of elementary channels. Each package comes with a set of interfaces and ports for accessing the channel. A designer of communication toolkits may provide additional packages based on the methodology that we will now describe. Or, he may provide additional channels that implement already existing interfaces—for instance, a signal with additional code to keep track of toggle statistics.

2.4.1 Interfaces

An *interface* consists of a set of operations. It specifies only the *signature* of each operation, namely, the operation's name, parameters, and return value. It neither specifies how the operations are implemented nor defines data fields, as there may be different implementations of the same interface.

In SystemC, all interfaces must be derived, directly or indirectly, from the abstract base class sc_interface. This base class provides a virtual function register_port(), which is called when a port is connected (via an interface, of course) to a channel. The arguments to this function are: (1) a reference to the port object, and (2) the type name of the interface that the port expects. The type name, which is derived from dynamic type information, enables the channel to ensure the legality of binding the port to the channel.

Let us first consider two simple interfaces used with the hardware signal: sc_signal_in_if<T> and sc_signal_inout_if<T>. The first of these interfaces is derived directly from sc_interface and is parameterized by data-type T. It provides a virtual function read() that returns a constant reference to T—so that the value may be only read. The second, sc_signal_inout_if<T>, provides a virtual function write() that takes as parameter a constant reference to T. Since sc_signal_inout_if<T> is derived from sc_signal_in_if<T>, it inherits from the latter the function read(). Thus, sc_signal_inout_if<T> consists of a read() and a write() function. The semantics of these functions are to read from and to write to a channel that implements the interfaces, though it is the channel that determines exactly how to carry out these operations. In SystemC, the definition of the *out*-interface is identical to the *inout*-interface. This is because oftentimes the value of a signal needs to be examined at an output port, e.g., for debugging purposes. The names of these interfaces and their associated port classes, however, are distinct so that the user can express his intentions clearly (see section 2.4.2, below).

Figure 2.2 shows the inheritance relationship of these interfaces, using the standard Object Modeling Technique (OMT) notation [31]. Now any channel that implements the sc_signal_inout_if<T> interface necessarily implements sc_signal_in_if<T> as well, and must provide the methods in both of them. Based on this interface derivation, we say that sc_signal_inout_if<T> is a *subtype* of sc_signal_in_if<T>. The notion of subtype is significant when we bind two ports corresponding to two interfaces thus related, as we will see in section 2.9.

2.4.2 Ports

Good design practice recommends that a model should interact with its environment, not arbitrarily, but through well-defined boundaries. Test-

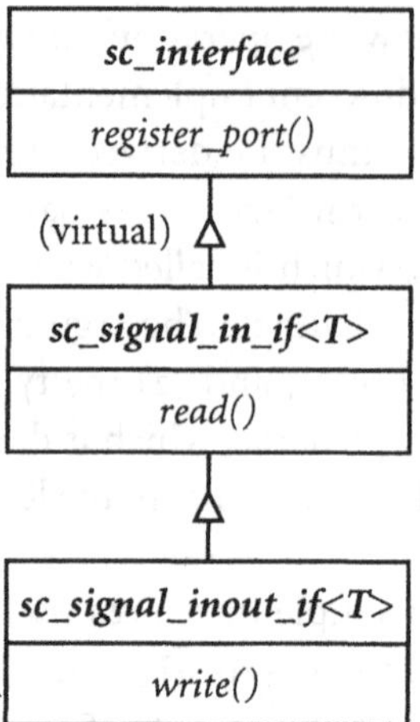

Figure 2.2: Class hierarchy of interfaces. *Italic* denotes abstract classes and virtual functions. Note that direct derivation from the base class `sc_interface` should be a virtual one so that we obtain the desired "diamond-shaped" inheritance [33] (not exhibited by this example) when we create composite interfaces.

benches may be an exception because they require much more flexibility and are not intended to be synthesized or refined into actual designs. But even so, it is always advisable to avoid implicit interfaces (e.g., sharing a global variable) whenever possible. Explicit interfaces greatly increase the reusability of the components we build.

To this end, SystemC provides *ports* for modules to connect to and communicate with their surroundings. We represent ports by objects, not simple pointers or references. This not only makes code more readable, but also allows the SystemC kernel to manipulate them in a way that simplifies the overall syntax for the user. With port objects, we do not require that ports be bound at module instantiation time and be repeated as formal parameters to the module constructor. This benefit comes with no performance overhead (when compared against pointers or references) in any reasonable implementation. The only, minor disadvantage is that correctness checking of port–channel bindings during module instantiation cannot be done at compile time (if the positional form is used; see below), but needs to be deferred to elaboration time.

Ports correspond to interfaces—each kind of port must specify the interface to which it corresponds. As with interfaces, the SystemC kernel interacts with port objects through an abstract class, `sc_port_base`, from which all ports are derived, directly or indirectly. Moreover, SystemC provides a template class for creating ports:

```
template<class IF, int N=1>
class sc_port : ... // class derivation details omitted
{
public:
    IF* operator->();
    // other member functions and member variables
};
```

The template class `sc_port` takes two parameters: an interface `IF` through which the port may be connected, and an optional integer `N` that signifies the maximum number of interfaces that may be attached to this port. The default value of `N` is one; when `N` is greater than one, we think of the port as a port array or *multiport*. If `N` is zero, an arbitrary number of channels may be connected to the port. For the present discussion we shall focus on simple ports (`N` = 1).

The most prominent method of `sc_port` is `operator->()`, which returns a pointer to the interface with which this port is associated. By virtue of C++ syntax we can invoke any method of `IF` directly from `sc_port`. Consider the following declaration within an `SC_MODULE`:

```
    sc_port<sc_signal_inout_if<int> > p;
```

Here we declare a port `p`, which can access a channel through the interface `sc_signal_inout_if<int>`. As we have seen in section 2.4.1, this interface provides the `read()` and `write()` methods. Thus, we may read the value of the channel and write to it using the expressions `p->read()` and `p->write()`. These are examples of *interface method calls* (IMC)—the channel's methods are invoked through an interface. IMC plays an important role in the design and refinement of communication protocols, as we will see in chapter 9.

Although the template class `sc_port` provides a systematic method for creating ports, it by no means restricts the functionality of the ports that may be designed. Often, the designer of a toolkit may wish to include additional features or syntax-simplifying wrappers ("syntactic sugars") for his ports. He can easily accomplish this by deriving the specialized port from `sc_port`, and add the desired features or syntactic sugars in the derived port. Indeed, the elementary channel package of SystemC defines the following ports, `sc_in` and `sc_inout` (called *signal ports*):

```
template<class T>
    class sc_in   : public sc_port<sc_signal_in_if<T> > ...;

template<class T>
    class sc_inout : public sc_port<sc_signal_inout_if<T> > ...;
```

The port class sc_out is identical to sc_inout. With these, let us continue with the example from section 2.3 by filling in the ports:

```
SC_MODULE(Adder) {
    sc_in<int>    a;
    sc_in<int>    b;
    sc_out<int>   c;
    // processes, etc.
    SC_CTOR(Adder) {
        // body of constructor:
        // process declaration, sensitivities, etc.
    }
};
```

We have now added to the module Adder two input ports (a and b) and one output port (c), all of data-type int. It is customary (and desirable, too) to declare the ports of a module first, for the ports define how the outside world views the module. Also, as we will see, the order of port declaration is significant when the positional form of port binding is used.

Ports of type sc_out and sc_inout can be used to initialize the signals to which they will be bound during elaboration, by way of the initialize(). Initialization of signals is important for ensuring deterministic simulation, because without initialization a signal may hold some unknown random value. Initialization should be placed within the constructor. For instance, within the constructor for Adder, we might initialize the value of the signal that will be bound to port c as follows:

```
c.initialize(0);
```

Afterwards, during elaboration, the value 0 will be assigned to the signal when the binding of c to the signal takes place. We will see more examples of signal initialization later in section 7.5.

2.4.3 Channels

Whereas interfaces and ports together describe what functions are available in a communications package, channels define how these functions are performed. We initiate operations through interfaces, but it is the channels that carry out these operations. We say that a channel *implements* an interface if it provides concrete definitions of all of the interface's operations and is properly derived from that interface.

Different channels may implement the same interface in different ways. On the other hand, a channel may implement more than one interface—provided it implements the operations specified in all of its interfaces. For instance, the hardware signal `sc_signal<T>` implements both the *in*- and *inout*-interfaces; "diamond-shaped" composite interfaces are also possible. Other than some methodological rules to which a channel must conform, SystemC places few restrictions on the functionality of channels. Thus, channels may vary widely in complexity—from hardware signals to complex protocols with embedded processes.

Within the context of a system model, channels provide means of communication between modules and between processes within a module. A channel, however, need not be limited to a point-to-point connection. Restrictions on connectivity and accessibility depend entirely on the channel itself. For instance, a hardware signal may have one writer but many readers, whereas a FIFO buffer must have exactly one input port and one output port connected to it. We will now take a detailed look at several classes of channels.

Primitive Channels

A *primitive channel* is one that supports the *request–update* method of access. The request–update method of access is closely related to the two-phase execution semantics of SystemC simulation (section 2.10), and is designed for simulating concurrency. When simultaneous actions have to be serialized (e.g., simultaneous read and write to a signal) or any form of arbitration is necessary (e.g., simultaneous bus requests), we need to delay any change to the internal state of the channel that might give rise to indeterminacy. That is to say, the result of simulation should not depend on the order in which simultaneous actions are executed. Thus, any operation that may potentially change the state of the channel (such as writing to it) will not have effect until after all currently active processes are finished or have come to a synchronization point. (At this point, the reader may wish to glance through section 2.10 to gain a first-order understanding of SystemC's simulation semantics.)

SystemC provides a base class called `sc_prim_channel`, from which all primitive channels are to be derived. Two methods of this class are germane to the present discussion: `request_update()` and `update()`. The first instructs the scheduler to place the channel in an update queue. Then, during the update phase of simulation, the scheduler takes items from the update queue and calls `update()` on each of them. Because its behavior is identical for all implementations of the interface, `request_update()` is defined as a nonvirtual function. In contrast, `update()` is virtual and each implementation is responsible for specifying what it actually does. Also, direct calls to `request_update()` are restricted to member functions of the channel only and should not be attempted by a process or a port.

We give a few examples of primitive channels below:

1. The hardware signal, `sc_signal<T>`, which implements the interface `sc_signal_inout_if<T>`. The semantics of the hardware signal are similar to the VHDL signal.

2. The FIFO channel, `sc_fifo<T>`, which implements the interfaces `sc_fifo_in_if<T>` and `sc_fifo_out_if<T>`. These interfaces provide both blocking and nonblocking versions of access. If the FIFO is empty, a blocking read will cause the calling process to suspend until more data is available, but the nonblocking version will simply do nothing. Likewise, if the FIFO is full, a blocking write will cause the calling process to suspend until more space is available. The actual insertion into or deletion from the FIFO takes place during the *update* phase of simulation, when the FIFO's `update()` function is invoked.

3. The mutual-exclusion lock (commonly called the *mutex*), `sc_mutex`, which is useful for modeling critical sections for accessing shared variables. A process, before entering a critical section, attempts to lock the mutex. If the mutex has already been locked by another process, then the mutex will cause the current process to suspend. By way of this example we point out that the use of `request_update()` and `update()` is optional. If the channel does not need a separate update phase, it does not call `request_update()` and may leave out the definition of its own `update()`.

Hierarchical Channels

A primitive channel, as the name implies, is atomic in that it does not contain other SystemC structures. Even when the behavior of a primitive channel involves thread blocking (e.g., in `sc_fifo<T>` and `sc_mutex`), the actions apply to the calling processes, which are conceptually external to the channel. The expressiveness of primitive channels could thus be limited to the simpler communication mechanisms only.

In contrast, hierarchical channels offer a more powerful method for modeling complex communication structures. For instance, the on-chip bus (OCB), which is the present standard system-on-chip (SoC) backbone from the Virtual Socket Interface Alliance (VSIA) [38], consists of several intelligent units, such as an arbiter, a control programming unit, and a decoder unit. This ensures full scalability, IP reuse, and a rapid time-to-market for SoCs. Primitive channels are not well suited for modeling such structures because of lack of internal processes. In this case, we need hierarchical channels, which, as modules that implement one or more interfaces, can have internal processes. We will see hierarchical channels in more detail in chapter 7 and chapter 8.

Hierarchical channels are also useful for the refinement of primitive channels. By refinement we mean the addition of more details to the model. Often, the first step in modeling is to write the model in an abstract, time-independent way, and then gradually add timing and structural details for more accurate analysis and estimation. For instance, we may refine the FIFO buffer in the previous section by explicitly including the handshakes that take place during reading and writing. The handshaking may be encapsulated inside the the `read()` and `write()` functions of a hierarchical channel `hw_fifo_wrapper<T>`, which implements the same interfaces as `sc_fifo<T>` (see section 7.5).

The advantages of using ports and interfaces to access a channel are now clear. With interfaces, we can easily plug in different versions of a channel into a model. In this example, `hw_fifo_wrapper<T>` and `sc_fifo<T>` implement the same interfaces in vastly different ways. Yet we can still substitute one for the other without having to worry about the details of modules that use them. "Designing to interfaces" has been a fundamental method in object-oriented software construction, and is increasingly being applied to system design as well. See [30] for an excellent overview of interface-based design.

2.4.4 Graphical Notations

Figure 2.3(a) shows the graphical notations that will be used in illustrations in the coming chapters. A module is simply shown as a box. An interface is represented by a circle with a "U-turn" arrow. A port is represented by a square with two arrows in opposite directions. A primitive channel is simply represented by a thick line, annotated with the name or type of the

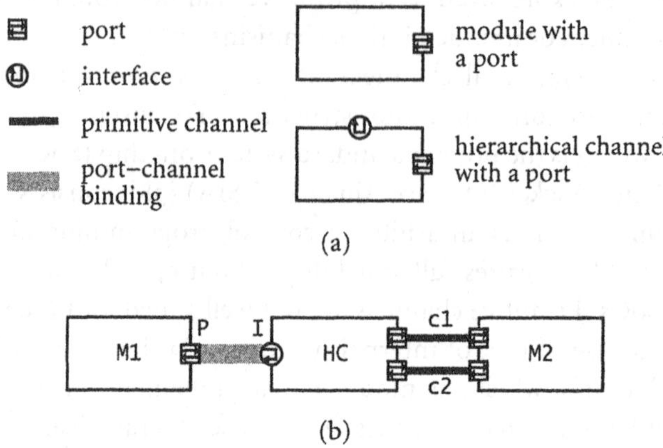

Figure 2.3: (a) Graphical notations for modules, interfaces, ports, channels, and port–channel binding (b) Example with two modules and a hierarchical channel

channel. In contrast, a hierarchical channel is represented by a box (not necessarily rectangular), like a module, with interfaces, and possibly ports as well, attached to the box. The binding of a port to a hierarchical channel is shown as a gray "pipe" that links the port to the interface through which the port is bound to the channel. The example shown in figure 2.3(b) consists of two modules, `M1` and `M2`, and a hierarchical channel `HC`. The port `P` of `M1` is bound to `HC` through interface `I`. `HC` itself has two ports, which are connected to the ports of `M2` via primitive channels `C1` and `C2`.

2.5 Processes

In SystemC the basic unit of functionality is called a *process*. In a typical programming language, functions are executed sequentially as control is transferred from one function to another. This is useful for modeling sequential behavior of systems. However, electronic systems are inherently parallel with many activities taking place concurrently. Processes provide the mechanism for simulating concurrent behavior.

A process must be contained in a module—it is defined as a member function of the module and declared to be a SystemC process in the module's constructor. Note that a member function in a module is not necessarily a SystemC process. A declaration in the module's constructor is required to register the member function with the simulation kernel. Processes access external channels through the ports of its containing module.

To accommodate the different needs in expressiveness and simulation performance, SystemC has two kinds of processes: *method process* and *thread process.* The macros SC_METHOD and SC_THREAD are used to make member functions behave as processes. These macros are quite complex and may even be dependent on compilers and platforms, as some compilers may not yet fully support certain C++ features. Therefore, we strongly recommend that the designer use these macros instead of spelling them out.

Continuing with the Adder example, let us add a member function and declare it to be a method process:

```
SC_MODULE(Adder) {
    sc_in<int>    a;
    sc_in<int>    b;
    sc_out<int>   c;

    void compute() {
        c = a + b;
    }
    SC_CTOR(Adder) {
        SC_METHOD(compute);
        sensitive << a << b;
    }
};
```

The member function compute(), when invoked, computes the sum of the inputs a and b, and writes the result to the output c. But the presence of a member function by itself does not create a process. Rather, inside the module's constructor, which is declared with SC_CTOR(Adder), the statement SC_METHOD(compute) maps the member function to a method process by registering it with the scheduler. The next statement specifies that this process is sensitive to changes in the values of hardware signals that will be connected to the input ports. Thus, whenever a or b changes, the value of c will be recomputed. (Section 2.7 describes sensitivity in more detail.)

A method process, when triggered, always executes its body from the beginning to the end. That is, it does not keep an *implicit execution state.* If the intent of the design requires different paths through the body of the process depending on some state information, the state information must be explicitly represented, usually as member variables in the module. For instance, a common way to model a finite state machine using SC_METHOD is to use a member variable to represent the state explicitly, and use a switch statement to determine the actions taken according to the state and the inputs. We will consider detailed examples in chapter 4.

A thread process, on the other hand, may have its execution suspended by calling the library function wait() or any of its variants. The thread remembers the point of suspension, as it were, along with all local variables, so that when the execution is resumed, it will continue from that point, rather than from the beginning of the process. Thus, unlike method processes, a thread process implicitly keeps its state of execution. This feature allows for far greater expressiveness for thread processes than method processes. For example, by means of wait statements multicycle behavior may be easily described in SC_THREAD, but would require more effort with SC_METHOD.

The gain in expressiveness, however, comes with a cost in simulation performance when the reference implementation of SystemC and a standard C++ compiler are employed. As the semantics of process suspension and resumption depart from traditional sequential software, we need *coroutines* [16] to implement thread processes. Coroutines require operating system and instruction set support, and switching between coroutines (known as a *context switch*) may be expensive on some microarchitectures. Nevertheless, since SC_THREAD is mainly used for modeling at the behavioral and higher levels in which the code between wait statements does substantially more computational work than an RTL cycle, and, as a result, the proportion of time spent in context switching is relatively lower, this trade-off may be acceptable in practice.

Even so, as simulators can never be too fast, there will always be demand for dedicated simulators. A dedicated simulator for SystemC can substantially improve the performance of coroutines (among other things) by optimizing the context switches beyond what a C++ compiler can do.

Like method processes, a thread process may have a sensitivity list describing the set of events to which it should normally react. As we have seen, when we encounter a wait statement, the execution of a thread process is suspended. When any of the events in the sensitivity list occurs, the scheduler will resume the execution of the process from the point of suspension. For example, a synchronous, edge-triggered design, when modeled with SC_THREAD, will have in its sensitivity list exactly one event: either the positive edge or the negative edge of a boolean signal that serves as its clock. Inside the body of this design, then, wait statements signify clock boundaries.

A thread process may also wait for events other than those specified in the sensitivity list. This is known as dynamic sensitivity, which we will discuss in section 2.7.2. In any case, when a thread process is resumed, it will run until it encounters a `wait` statement of some kind. It is thus possible, if the designer is not careful, for a thread process to monopolize the simulator and cause other processes to starve. A common mistake is to forget a `wait` statement inside a potential infinite loop. Likewise, a method process may also be caught unintentionally in an infinite loop. On the other hand, when a member function corresponding to a thread process returns, the thread process terminates. The designer should always keep these pitfalls in mind, as he does with any concurrent-programming activity (not just with SystemC).

It should also be noted that, in simulation, the concurrency among processes is usually only apparent, not actual. In other words, when faced with a choice of two or more processes that are triggered at the same time, the scheduler arbitrarily selects an order. This order is typically *deterministic* in that the scheduler either does not randomize it or randomizes it according to a deterministic (pseudorandom) generator. Hence, two simulation runs with exactly the same model (including order of declaration), same inputs, same version of kernel, same generator seed, etc., will produce the same results. But this order is *unspecified*, and the designer ought not depend on the order selected by a particular simulation engine. This implies, for example, that the use of global variables or shared member variables should be avoided. If two processes in a system communicate through a global variable or a shared member variable, the system may behave as desired with a particular simulation engine, yet may no longer do so when the simulation engine is modified but still complies with the SystemC specification.

2.6 Events

We have until now spoken of events in an intuitive sense. Now let us consider events more rigorously. An *event* is an object, represented by class `sc_event`, that determines whether and when a process's execution should be triggered or resumed. (Henceforth we will use triggering and resumption synonymously when we do not distinguish between methods and threads.) In more concrete terms, an event is used to represent a *condition* that may occur during the course of simulation and to control the triggering of processes accordingly. We need to distinguish an event from

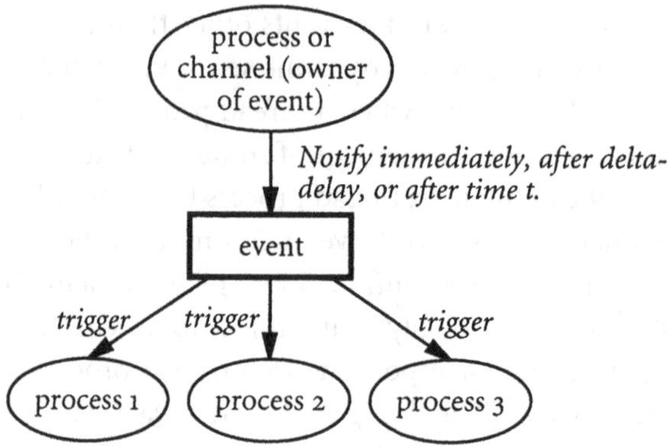

Figure 2.4: Event notification and process triggering.

the actual *occurrence* of an event. An event is, as it were, in the "subjunctive mood" (that is, expressing *contingency*, as in "*if* it *should* rain"), whereas an actual occurrence of it is in the "indicative mood" (as in "it's raining today"). There may be multiple occurrences of an event, and each occurrence is unique though reported through the same event object.

An event is usually, though not necessarily, associated with some change of state in a process or of a channel. This association, however, is transparent to the event object: the *owner* of the event is responsible for reporting the change to the event object. The act of reporting the change to the event is called *notification*. The event object, in turn, is responsible for keeping a list of processes that are *sensitive* to it. Thus, when notified, the event object will inform the scheduler of which processes to trigger. Figure 2.4 shows how an event object interacts with its owner and the scheduler.

In classical hardware modeling, a hardware signal has an event associated with the change in its value. The signal is responsible for notifying the event whenever its value changes. A signal of boolean type has two additional events, one associated with the positive edge and the other with the negative edge. A more complex channel, such as a FIFO buffer, may have an event associated with the change from being empty to having a word written to it, and another with the change from being full to having a word read from it.

An event object may also be used directly by one process P_1 to control another process P_2. If P_1 has access to event object E and P_2 is sensitive to or waiting on E, then P_1 may trigger the execution of P_2 by notifying E. In this case, event E is not associated with the change in a channel, but rather with the execution of some path in P_1.

As we have seen, an event notification will cause processes that are sensitive to it to be triggered. We have not, however, stated *when* the execution of the processes will take place. The timing of the notification is specified at the invocation of the `notify()` method and may be one of the following: immediate, delta-delay, or nonzero delay.

An *immediate notification* causes the sensitive processes to be triggered at the current time instant, but before updates in primitive channels take effect. In contrast, a *delta-delay notification* delays the triggering of the sensitive processes by an "infinitesimal" amount of time, during which channels are updated. Thus, for example, if a signal has had a new value written to it at the current time instant, processes that are triggered by immediate notification will still see the *old* value of the signal, whereas processes triggered by delta-delay notification will see the new value. Finally, a *nonzero-delay notification* puts the event in a queue, ordered by time, so that processes sensitive to the event will be triggered after the designated amount of time. It should be noted that, in our terminology, notification occurs at the time when an event is removed from the queue and its sensitive processes are triggered, not when the event is placed in the queue. Hence, we construe `e.notify(t)` as "notify event `e` after time `t`." We will revisit event notification in section 2.10 when we discuss SystemC's simulation semantics.

2.7 Sensitivity

In our example `Adder`, we specify in the constructor that the method process `compute` is sensitive to events on the input ports, as the syntax suggests. This is accomplished by passing the ports via the << operator to a special member variable of `sc_module` named `sensitive`:

```
sensitive << a << b;
```

Just like C++-style I/O, the << operator may be cascaded arbitrarily for `sensitive`. The sensitivity declaration applies to the process that is most recently declared. Thus, if there are two or more processes in a module, each process declaration must be followed immediately by its own sensitivity declaration.

2.7.1 Static Sensitivity

In `Adder`, the list of events to which `compute` should respond is determined before simulation begins. This list remains the same throughout simulation. We call this a *static sensitivity list.* Normally, RTL and synchronous behavioral processes use static sensitivity lists only (chapter 4).

2.7.2 Dynamic Sensitivity

It is possible for a process to temporarily override its (possibly empty) static sensitivity list. That is, during simulation a thread process may suspend itself and designate a specific event e as the current event on which the process wishes to wait. Then, only the notification of e will cause the thread process to be resumed; the static sensitivity list is ignored.

To wait for a specific event e, the thread process simply calls `wait()` with e as argument:

```
wait(e);
```

Composite events, for use with `wait` only, may be formed by conjunction (AND) or disjunction (OR):

```
wait(e1 & e2 & e3);
wait(e1 | e2 | e3);
```

The first of these two statements waits for all three events to be notified, in any order, whereas the second statement waits for any of the three events to be notified. (At the time of this writing, SystemC does not support a mixture of conjunction and disjunction.) The `wait()` function may also take as argument a time, e.g.,

```
wait(200, SC_NS);
```

or, equivalently,

```
sc_time t(200, SC_NS);
wait(t);
```

Here we have two equivalent forms of waiting for 200 nanoseconds to elapse. We accomplish this by creating a temporary event along with a notification scheduled at 200 ns later. If the time interval is zero, the effect is to wait for a delta-cycle.

Finally, by combining time and events, we may impose a timeout on the waiting of events. For instance,

```
wait(200, SC_NS, e);
```

waits for event e to occur, but if e does not occur within 200 ns, the thread process will give up on the wait and resume. Also, the process will remove itself from the list of processes waiting on e.

Dynamic sensitivity is also applicable to method processes. Instead of calling wait(), a method process calls next_trigger() to specify the event that must occur for it to be triggered next time. Until the event occurs, the static sensitivity list is temporarily disabled. Unlike wait(), however, calling next_trigger() does not suspend the current method process. Instead, execution of the method process continues to the end, and next time the method process will be invoked only when the event specified by next_trigger() occurs. If an invocation of a method process does not call next_trigger(), then the static sensitivity list will be restored.

The types of events that next_trigger() accepts are identical to those accepted by wait(). Thus, for example,

```
next_trigger(200, SC_NS, e);
```

will make the current method process wait on e within a timeout of 200 ns. If e occurs within 200 ns, the method process will be triggered. Otherwise, when the timeout expires, the method process will be triggered and its static sensitivity list will be back in effect. It should be noted that, if next_trigger() is called more than once during the execution of a method process, the last call supersedes all others.

Use of Dynamic Sensitivity

Let us now revisit the FIFO example of section 2.4.3. We described the conditions that can suspend a calling process. The suspension is achieved by waiting on events that indicate that those conditions no longer hold. For example, in the blocking read, we have:

```
while (num_available() == 0)
    wait(_data_written);
```

where _data_written is an event that is notified when some data is written into the buffer so that the read operation may proceed. Likewise, a blocking write waits on the event _data_read if the buffer is full. This eliminates the need for explicit synchronization and simplifies for the caller the job of getting data in and out.

Dynamic sensitivity is also a useful mechanism for controlling software tasks and their interaction with hardware processes. (Software tasks will be supported in future versions of SystemC.) In systems containing software tasks, processes may be created and later destroyed during execution time. For instance, a network server might spawn off new processes to handle network connection requests, and these processes are destroyed afterwards when the connections are terminated. Hardware processes that interact with the software tasks would be sensitized and desensitized accordingly.

*2.8 Event Finders

We recall that the declaration `sensitive << a << b;` adds the ports a and b in the static sensitivity list of the most recently declared process. More strictly speaking, however, what this declaration does is to make, at elaboration time, the process sensitive to some event associated with the channel that will be bound to the port. Here, because ports of the `sc_in` class correspond to an interface for hardware signals (`sc_signal_in_if`), we are able to infer that the event in question is the default event, namely, the event associated with a change in the value of the signal. Nevertheless, it is perfectly possible for a channel to have more than one kind of event (e.g., a FIFO channel). Since the sensitivity declaration takes place during module instantiation and *before* port–channel binding, what is given to `sensitive` cannot be an actual event object. Therefore, the processing of sensitivity must somehow be deferred until we know the channel that owns the desired event object.

Here we introduce the concept of event finder. An *event finder* is an object that is associated with a port P and a method M that returns an event object when invoked on a channel (through an interface). When P is bound to a channel C through interface I, the method M may be invoked on I to obtain the desired event object. Note that M is not a method of an event finder, but (in C++ terms) a member-function pointer that may point to member functions of the interface I. Event finders require careful implementation, as the simulation kernel does not know anything about the interface I other than that it is derived from `sc_interface`. One possible solution (among many, to be sure) is to define an abstract class `sc_event_finder`, which consists of a single virtual function `find()`, and to derive from it a template class parameterized by interface I. Having I as a template parameter allows us to construct the type of the member-function pointer safely. The kernel, after having connected the port to the channel, then calls `find()` on each event finder to obtain the desired event object.

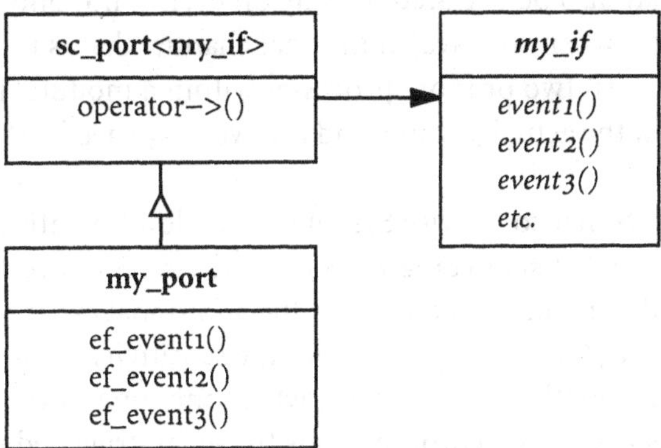

Figure 2.5: Event finders. The interface `my_if` has three events. The port class generated by the template `sc_port` provides `operator->()` for access to `my_if`'s methods. It is then specialized to `my_port`, which has an event-finder function for each of `my_if`'s events.

It is now easy to see that, for every event E visible through interface I, port P should provide an event finder, F_E. If we wish to make a process sensitive to event E, we will pass F_E to `sensitive`. For instance, if an event finder associated with an event `event1` can be obtained by calling `ef_event1()` on port `p`, we would say:

```
sensitive << p.ef_event1();
```

to make the most recently declared process sensitive to the event when port binding takes place.

The burden of providing event finders in port classes falls on the designer of the communications package. As we saw in section 2.4.2, SystemC supplies a template `sc_port<IF,N>` for generating port classes. Because of its generality, this template does not provide any functionality that depends on the specifics of interface `IF`—it simply forwards requests by way of `operator->()`. In particular, it does not provide event finders for the events accessible through `IF`. Hence, if event finders are required, we need to derive a port subclass, in the same manner as `sc_in` and `sc_inout` were derived in section 2.4.2, and add the event-finder functions to that subclass. Figure 2.5 shows such a port subclass, `my_port`, derived from the template instance `sc_port<my_if>`. For each of `my_if`'s events there is a corresponding event-finder function in `my_port`, with the `ef_` prefix chosen only as a convention.

Sensitivity may also be declared directly on events, not only event finders. This is useful when we have an internal channel that is used for communication between two or more processes within a module. In this case, because we know the actual channel and the event, we need not go through event finders.

The signal ports that come with SystemC provide automatic conversions to event finders, so that `sensitive` works with these ports directly. In addition, signals and ports of `bool` type have the additional event-finder functions `pos()` and `neg()`, so that one can declare sensitivity to either only the positive edge or only the negative edge when only one of the edges is of interest. For instance, a synchronous, positive-edge-triggered design may simply state:

```
sensitive << clk.pos();
```

where `clk` is a port of `bool` type. This may be used for both method and thread processes, and is commonly seen in classical RTL and behavioral descriptions (chapter 4).

In sum, event finders are only applicable to static sensitivity lists, since dynamic sensitivities always work on events directly. Even with static sensitivity, event finders are explicitly required only when we need to find a nondefault event through a port. For ports whose corresponding interfaces have merely default events, such as `sc_in`, we may simply use the name of the port in declaring sensitivity.

2.9 Module and Channel Instantiation

We have examined the modeling elements of SystemC, their functionality and interaction. We now turn to the construction of instances of modules and channels for simulation.

As in any object-oriented framework, a class definition describes the features of a family of objects, but does not create objects. Thus, for example, the definition `SC_MODULE(Adder) {...}` does not create adders, but merely says what an adder looks like.

Before we can simulate a system, we need to create its components, namely modules and channels, and connect them up with one another. This is called *instantiation*. Instantiation is not limited to the top level—a module or a hierarchical channel that contains other modules and channels needs to instantiate them, usually in the constructor. (At present we do not consider dynamic instantiation, a capability that will be supported in future generations of SystemC; we assume that the structure of the system is determined at elaboration time and does not change during simulation.) Then, when we instantiate a higher-level module, it will recursively instantiate its descendents.

2.9.1 Hierarchical Modules

Let us consider a module that consists solely of instances of another module and an internal channel:

```
SC_MODULE(Add3) {
    sc_in<int>     in1;
    sc_in<int>     in2;
    sc_in<int>     in3;
    sc_out<int>    sum;

    sc_signal<int> temp;                 // internal signal

    Adder* adder1;
    Adder* adder2;

    SC_CTOR(Add3) {

        adder1 = new Adder("adder1");
        (*adder1)(in1, in2, temp);       // positional form

        adder2 = new Adder("adder2");
        adder2->a(temp);                 // named form
        adder2->c(sum);                  // order does not matter
        adder2->b(in3);
    }
};
```

Add3 describes a 3-input adder that is structurally composed of two 2-input adders. After the input and output ports, we declare a temporary signal, temp, and two Adder pointers, adder1 and adder2. In the constructor, we instantiate the temporary signal (implicitly done by the C++ compiler) and the two 2-input adders.

We give each of the adders a string name. The string name need not always match the instance name, but we will not deliberately confuse ourselves unless unusual circumstances require a different name. The string name allows the SystemC library to assign a hierarchical name to the instance automatically. This hierarchical name would be formed by the concatenation of the parent's hierarchical name and the string name just given.

For example, if an instance of `Add3` is named `"abc"`, then the hierarchical name of its `adder1` submodule will be `"abc.adder1"`.

This example also illustrates the two forms of port binding. The first, the *positional form*, relies on the order in which the ports are declared in the definition of the module. Although this form is more concise, we cannot check at compile-time the correctness of the number of signals and of correspondence of types between the ports and the signals. A more verbose alternative is the *named form* of binding, which we use to bind the ports of `adder2`. With this form, the order of binding does not make a difference, and some type checking can be done at compile-time (e.g., a port of data-type `int` must not be bound to a signal of data-type `double`). In either case, however, only during elaboration can SystemC discover any unbound ports and warn the designer about them.

Finally, we should observe that a port may be bound to either a channel or another port. In the case of `adder1`, `a` and `b` are bound to ports `in1` and `in2`, and `c` is bound to `temp`, a channel (in this case, a signal). Binding a port P_1 to another port P_2 is meaningful only if P_2 is a port of the parent module. What this binding indicates is that, at elaboration time, P_1 will be eventually bound to the channel to which P_2 is bound. In addition, P_2's interface must be the same as, or a *subtype* of, P_1's interface. (Note that a subtype's interface is a *superset* of that of its supertype.) For example, if P_1 implements the interface `sc_signal_in_if<T>`, the binding is legal if P_2 implements `sc_signal_in_if<T>` or `sc_signal_inout_if<T>`. On the other hand, if P_1 implements `sc_signal_inout_if<T>` and P_2 implements `sc_signal_in_if<T>`, then the binding is illegal—since P_2's interface determines what operations are available inside the parent module. In this case, P_2 does not support writing to the channel; therefore, neither should P_1.

2.9.2 Top-Level Instantiation

At the top level, we instantiate the modules to be tested, the testbench, and the channels that connect them. It is, in fact, desirable to group them all together in a module and just to instantiate the resulting module, so that the model may be cleanly separated from command-line processing and simulation control. But in any case we need to be mindful of the lifetimes of these objects—if any simulation object goes out of scope before or during simulation, the simulator will not behave as expected. Hence, if objects are instantiated locally in a function, we need to start simulation in that function as well, e.g.,

```
void toplevel()
{
    sc_signal<int> sig_a, sig_b, sig_c;
    Adder my_adder("my_adder");
    my_adder(sig_a, sig_b, sig_c);

    // Other modules and testbench, which drive sig_a and sig_b

    // Start elaboration and then simulation
    // Run for 1000 seconds of simulation time
    sc_start(1000, SC_SEC);
}
```

Here, `sc_start(1000, SC_SEC)` runs the simulation for 1000 seconds of simulation time. When `sc_start()` returns, we may continue the simulation from where it stopped by calling `sc_start()` again. But the modules and channels are required to be still alive—in this case, any subsequent call to `sc_start()` must lie within the function `toplevel()` because the modules and signals, declared as objects within the function, will cease to exist when `toplevel()` exits.

The entrance from the SystemC library to the designer's code is through the function `sc_main()`, which receives the standard command-line arguments `argc` (of type `int`) and `argv` (of type `char**`). This function typically consists of initialization, simulation (e.g., calling `toplevel()` above), clean-up, and returning a status code. It is the responsibility of the designer to define this function. In some cases the `main()` function provided by the SystemC library may not meet the user's need. The user will then have to mimic SystemC's `main()` (in file `sc_main.cpp`) and include the new `main()` within his design—this will cause the linker not to link in the `sc_main.o` from the SystemC library file.

2.10 Simulation Semantics

The scheduler is the heart of SystemC; it controls the timing and order of the execution of the processes, handles event notifications, and updates the channels that request to be updated. The most intricate part of the scheduler pertains to the simulation of conceptually concurrent actions on a computer that has a smaller degree of parallelism.

Like VHDL and Verilog, the SystemC scheduler supports the notion of delta-cycles [23]. A delta-cycle lasts for an infinitesimal amount of time and is used to impose a partial order of simultaneous actions. It is a common device for interpreting zero-delay semantics. Thus, when the scheduler processes a delta-cycle, it does not advance the simulation time, but it executes actions that are scheduled at the current time in two phases: the *evaluate* phase and the *update* phase. As we have alluded to in section 2.4.3, the two-phase semantics allows simultaneous accesses to certain channels, such as signals, to be correctly simulated. (We say that two actions are *simultaneous* if they occur within the same simulation phase and either action is free to execute before the other.) Also, the coroutines in SystemC are nonpreemptive. This means that, for thread processes, code delimited by two `wait` statements always execute without any other thread processes' interruption.

The workings of the SystemC scheduler may be summarized as follows:

1. *Initialize.* During initialization, each process is executed once (for `SC_METHOD`) or until a synchronization point (i.e., a `wait`) is reached (for `SC_THREAD`). In some circumstances it may not be desired for all processes to be executed in this phase. To turn off initialization for a process, we may call `dont_initialize()` after its `SC_METHOD` or `SC_THREAD` declaration inside the constructor. The order in which these processes are executed is unspecified, though deterministic. This means that, even though a different version of simulator may choose a different order, two simulation runs using the same version of simulator yield identical results.

2. *Evaluate.* Select a process that is ready to run and resume its execution. This may cause immediate event notifications to occur, possibly resulting in additional processes being made ready to run in this same phase. Here we advise the reader to be cautious when using immediate notification. If a process *P* has already been executed in the current evaluate phase, it will be triggered again if a controlling event is notified immediately. On the other hand, if *P* is still to be executed in the current phase, then it will be triggered only once. Hence, if not used carefully, immediate notification may cause processes to be triggered an indeterminate number of times in the same evaluate phase—an undesirable behavior.

3. If there are still processes ready to run, repeat step 2.

4. *Update.* Execute any pending calls to `update()` resulting from calls to `request_update()` made in step 2. Recall that a primitive channel uses `request_update()` to have the scheduler call its `update()` function after the execution of processes. This is the phase when these

function calls take place. The updating of channels could generate notifications of events, and some or all of these notifications may be of the delta-delay type.

5. If there are pending delta-delay notifications, determine which processes are ready to run due to the delta-delay notifications and go to step 2. We do not advance simulation time when processing delta-cycles.

6. If there are no more timed (nonzero-delay) notifications, simulation is finished.

7. Otherwise, advance the current simulation time to the earliest pending timed notification.

8. Determine which processes are ready to run due to the events that have pending notifications at the current time. Go to step 2.

The scheduler, once started, continues indefinitely until there are no more events or a process explicitly stops it (or an exceptional condition occurs, such as writing simulation output to a full disk). Alternatively, we may invoke the scheduler by indicating the amount of time to simulate, as in the `toplevel()` example of section 2.9.2. Once the scheduler returns, we may continue the simulation from the time when the scheduler last stopped. This is a useful interface when a SystemC model serves as a "slave" process for a "master" process that controls system simulation.

2.11 Summary

We have presented the fundamentals of SystemC, the core language that lays the foundation upon which to build system-level models. SystemC provides a simulation kernel whose semantics are similar to traditional event-based simulators. Yet this kernel is not limited to simulation of classical hardware, though it is perfectly capable of it.

Most significantly, by carefully making abstractions for processes and channels, and using events as the mechanism of interaction between them, we decouple the simulation kernel from the formerly lower-level semantics of these entities. As we have seen from the examples in this chapter, this decoupling allows us to model, at various levels of abstraction, complex structures and behaviors such as busses and buffers. Although we have only looked at a few examples with a quick glance, the benefits of abstractions in SystemC will become clear in the chapters to come. We will also attain a deeper understanding of how SystemC enables the introduction of new models of computation and the refinement of system specifications to mixed software and hardware implementation.

3

Models of Computation

3.1 Introduction

SystemC allows design teams to build models of systems that utilize one or a mixture of various models of computation.

The notion of a "model of computation" (MOC) is fundamental to system level design; yet it has not been formalized completely. One attempt at a formal definition is found in [10], where the authors state: "How the abstract machine in an operational semantics [...of a language ...] can behave is a feature of what we call the *model of computation* underlying the language." The authors further discuss features of the model of computation, including the kinds of relations that are possible in a denotational semantics, the communication style, how individual behaviors are aggregated to make complex compositions, how concurrency is handled, and how such compositions are hierarchically abstracted.

For our purposes a less formal definition will suffice. We define the model of computation provided by a modeling language as follows:

1. The model of time employed (real-valued, integer-valued, untimed) and the event ordering constraints within the system (globally ordered, partially ordered, etc.).

2. The supported method(s) of communication between concurrent processes.

3. The rules for process activation.

Most "traditional" design languages such as VHDL, Verilog, and SDL can be seen as having a single fixed model of computation, and provide little or no way for users to customize the given model of computation. In a sense, SystemC also has a single fixed model of computation, but it is different from other traditional design languages in several key ways:

1. The "base" model of computation is designed to be extremely general.

2. The SystemC language as a whole is designed so that customized models of computation can be cleanly and efficiently layered on top of the base capabilities provided by the SystemC core language.

Briefly, the SystemC core language features that allow customized models of computation to be constructed include:

- *Events, the `notify()` call, and the `wait()` call.* These simple, flexible synchronization capabilities allow a broad range of different channel types to be implemented without having to change the underlying simulation engine. All the required functionality is already present in the simulation kernel. While the global model of time is fixed to an integer model, designers can construct specific channels to achieve their precise rules for communication between processes, process activation, and systemwide event ordering. It is important to realize that events are the *fundamental* synchronization mechanism within SystemC, and *all* channels that provide communication and synchronization (even ones as basic as `sc_signal`) are built up from events and other core language features.

- *The user's ability to create specialized channels, interfaces, ports, and modules.* SystemC allows users to create channels and interfaces to model specific communication capabilities in the same way that specific communication capabilities are provided by the `sc_signal` and `sc_fifo` channels introduced in chapter 2. SystemC also allows users to create specialized ports and modules that make using such user-defined channels more natural and less error-prone. As a simple example of this, a user could create a specialized module named `sc_rtl_module` that checks to make sure that all of its ports connect to signal channels. Users might then create their RTL modules by deriving from `sc_rtl_module` rather than the more general `sc_module`.

In this way, SystemC cleanly supports the emulation of a large number of different MOCs. Although continuous time models as used, for example, in analog modeling, cannot yet be constructed in SystemC, virtually any discrete time system can be modeled in SystemC. Some well-known models of computation which can be quite naturally modeled in SystemC include:

- Static multirate dataflow
- Dynamic multirate dataflow
- Kahn process networks
- Discrete event as used for
    - RTL hardware modeling
    - network modeling (e.g., stochastic or "waiting room" models)
    - transaction-based SoC platform modeling

Note that the above list of MOCs that can be modeled within SystemC is in no way exhaustive.

Since SystemC can support multiple MOCs, it is important to know how two or more *different* MOCs will interact when used in the same system. Here again SystemC's general-purpose, flexible core language features provides many benefits. In the same way that users can create specialized channels to precisely control communication within a single MOC, they also can create specialized channels to precisely control communication and synchronization between two or more MOCs. As a simple example, when a dataflow module sends tokens to an RTL module, it might be desirable to have the dataflow tokens be spread in time in some fashion (e.g., one token per clock edge). Users can construct MOC "conversion" channels that allow different MOCs to be interfaced according to their precise rules. We will see an example of such an MOC conversion block in chapter 10, example 10.2.3.

Why is it necessary for SystemC to support many different models of computation? One primary reason is that many systems are not homogeneous, and thus they are best described using a combination of models or notations which are natural to various subsystems.

For example, in figure 3.1, we depict a CDMA transmitter/receiver system. It consists of several key subsystems, which would be implemented in the real-world in many different design forms: Radio-frequency (RF) processing, mixed-signal blocks (D/A and A/D), digital hardware implementing DSP-style functions (e.g., channel coding, Viterbi decoding), possibly software implemented on a digital signal processor (DSP), and finally control software implemented on a standard microprocessor or microcontroller.

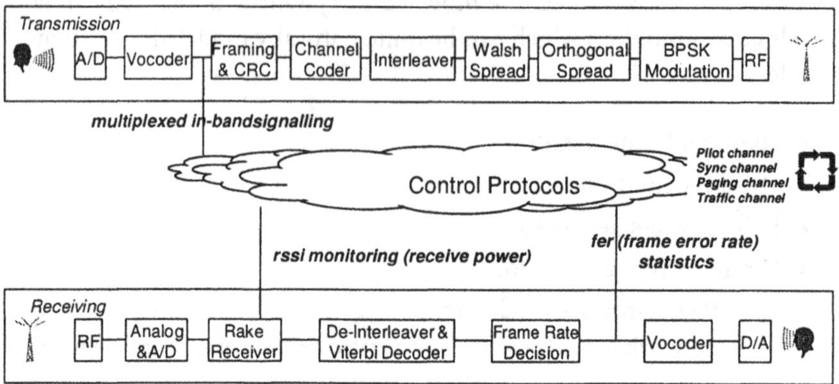

Figure 3.1: CDMA transmitter/receiver

The "natural" notation or model of computation for these various subsystems differs. The DSP-style dataflow functions might best be modeled initially as static or dynamic multirate dataflow. RF and mixed-signal blocks might be modeled using special continuous-time, differential equation based modeling representations. Control software could be modeled in a discrete event form, perhaps using transaction-level models. An underlying instruction-set simulator (ISS) for the control processor or a DSP might use an RTL MOC. In other words, the "natural" way to model this entire system is not to choose one particular model of computation and force every subsystem into it: rather, it is to build an overall system model composed from a series of heterogeneous subsystem models. This requires a modeling environment such as SystemC that allows the definition of models using particular models of computation, and a methodology for composition which allows one to make sense of the overall combination.

Another primary reason that it is important to support multiple MOCs within SystemC is that a particular subsystem might be most naturally expressed with different MOCs at different parts of the design refinement process. For example, a block that is most naturally expressed using static multirate dataflow at the functional level might be remodeled using an RTL MOC as it is refined to hardware.

The rest of this chapter briefly discusses several important models of computation that are commonly used in system design. Our intent is to provide the reader with a basic understanding of these models of computation and some idea of how they can be modeled within SystemC. Keep in mind that many other MOCs can also be modeled in SystemC, and that a given MOC may be modeled in many different ways in SystemC. The approaches outlined below for modeling MOCs in SystemC are for illustrative purposes only. Other more sophisticated approaches might be preferred in some cases for higher performance or greater flexibility.

It is also important to note that there currently are several working groups within the Open SystemC Initiative focused on developing standards for expressing various models of computation in SystemC, including a working group focused on analog and mixed signal modeling within SystemC.

3.2 The "RTL" Model of Computation

The register-transfer level (RTL) MOC is an informal term for a modeling style that corresponds to digital hardware synchronized by clock signals. This modeling style is widely used within languages such as Verilog and VHDL, and it is widely supported by commercial hardware synthesis tools. This modeling style will be described in detail in chapter 4.

In the RTL style, all communication between processes occurs through signals (`sc_signal`, `sc_signal_rv`, etc.). (The inner workings of SystemC signals will be described in detail in chapter 7.) Processes may either represent sequential logic, in which case they are sensitive to a clock edge, or they may represent combinational logic, in which case they will be sensitive to all of their inputs. RTL modules are *pin-accurate*. This means that the ports of an RTL module directly correspond to wires in the real-world implementation of the module. RTL modules are also cycle-accurate.

SystemC signals closely mirror the behavior of VHDL signals, and also the behavior of Verilog signals in which the Verilog deferred assignment operator (<=) is used. The signal channels provided within the SystemC release do not allow time delays to be specified when signal assignments are performed, although it would be easy to create a customized signal channel that provides this feature. (In general, when performing RTL modeling, time delays for signal assignments are not needed because combinational logic delays are not being modeled.)

Because combinational or gate delays are generally not modeled in SystemC at the RTL level, some people might mistakenly believe that SystemC is an RTL "cycle simulator". In true cycle simulators, the order of evaluation of processes is determined before simulation starts using either explicit manual coding or automatic techniques. In SystemC, the order of process evaluation is dynamically (and automatically) determined as the simulation proceeds.

3.3 Kahn Process Networks

A Kahn process network (KPN) is an effective model of computation for building algorithmic models of signal-processing applications. Such applications are common in the multimedia and communications product domains. The KPN model has been used for multimedia applications by Philips with its YAPI approach [8]. Also see [14] for more information.

In the KPN model, computing blocks or processes execute concurrently, and are connected by channels that carry sequences of data tokens. These channels are infinite length FIFO channels. The concurrent processes that model system functionality read and write to these FIFO channels with blocking read and nonblocking write operations. Systems modeled exclusively using KPN are deterministic: the process execution order (i.e., the scheduling order) will not affect the token values seen on the FIFO channels.

In chapter 5 we will see how `sc_fifo` channels and `SC_THREAD` processes can be used to easily model KPN designs within SystemC. As discussed in chapter 5, practical implementations of KPN systems cannot have infinite length FIFOs. Either specific upper bounds for FIFOs must be specified by the designer (thus making write operations potentially blocking), and/or a KPN scheduler must be provided that attempts to keep the network executing without letting any of the FIFOs grow to infinite length.

In its purest form KPN systems have no concept of time. However, as we will see in chapter 5, after initial functional modeling is completed it is often useful to annotate time delays (using `wait(sc_time)` statements) into KPN models. The resulting model is called a *timed functional model*, or also a *performance model.*

3.4 Static Dataflow

Static dataflow (SDF) networks are a special case of Kahn process networks in which:

1. The functionality within each process is clearly separated into three stages: reading of all input tokens, execution of the computation within the process, and writing of all output tokens.

2. The number of tokens that a process will read and write each time it is executed is fixed and known at compile-time.

Because the activation rules for each process are known at compile-time, tools can analyze the network and build static execution schedules for processes and compute bounds for all FIFOs at compile-time rather than at execution time. The resulting models will execute far more quickly than KPN or discrete event models due to the avoidance of dynamic scheduling. Although the scope and expressiveness of SDF is more limited than KPN or other discrete event MOCs, many algorithms have successfully been modeled using SDF. See [19] for further information on SDF. A good summary of methods for determining valid schedules for SDF networks that are consistent and deadlock-free can be found in [3].

One way to express SDF models within SystemC is to use the dataflow modeling style presented in chapter 5, "Functional Modeling." Keep in mind that the dataflow modeling style presented in chapter 5 is really the more general KPN MOC, since the token production and consumption rates for blocks are not explicitly declared by the designer, and dynamic scheduling is still used when the system is simulated rather than more efficient static scheduling.

3.5 Transaction-Level Models

Transaction-level models represent one specific type of the discrete-event MOC. In transaction-level modeling (TLM), we model communication between modules using function calls that represent the transactions, typically supported by a target platform. Transactions can be viewed as having a specific start time, end time, and payload data.

Within TLM designs, `sc_signal` channels are usually avoided entirely—instead data is exchanged between different processes by reading and writing shared data variables. In the absence of a well-planned overall synchronization scheme, having multiple concurrent processes reading and writing to a common set of shared data variables can lead to unpredictable system execution. In chapter 8 we will present a detailed example of a TLM design and see how a *two-phase synchronization scheme* is used to insure predictable and deterministic system execution.

TLM designs are usually more concise and simulate much faster than corresponding RTL designs.

3.6 Summary

Models of computation are fundamental constructs for the modeling of complex heterogeneous systems at various levels of abstraction. SystemC has been defined and developed to support the flexible modeling capabilities required to model a wide variety of models of computation. Future chapters indicate how this flexibility can be used by designers.

4

Classical Hardware Modeling in SystemC

4.1 Introduction

As we climb up the ladder of abstraction levels toward the system level, the lower rungs become more and more "classical" in comparison. Hence, by *classical hardware modeling* we mean modeling hardware at the register-transfer level (RTL) and the behavioral level. (Here, behavioral level is just one notch up from RTL. In behavioral-level descriptions, inputs and outputs are synchronized by a clock, but the exact number of clock cycles is flexible. This approximately corresponds to the level of modeling for architectural synthesis in chapter 4 of [25]. We will elaborate on this point in section 4.3.) By no means does this imply that these levels are obsolete; rather, we emphasize the novelty of system-level modeling methodologies and the versatility of SystemC in spanning many abstraction levels, current and new.

To be sure, there are many aspects of abstraction, such as timing, function, and structure, and it is not necessarily strictly correct to say that one abstraction methodology is "higher" than another. Nevertheless, the modeling methodologies we review in this chapter pertain to details (e.g., pin-level accurate structures) that are often not fully described by other methodologies that we shall introduce later, and it is convenient to think of the present ones as being at lower levels.

Why is it necessary, in the context of system-level design, to allow for modeling hardware at the RTL and behavioral levels? Because a large system is usually composed of many components that have varying requirements on performance, area, ease and flexibility of implementation, and other measures of cost, we need a broad range of abstraction levels to manage these requirements and trade-offs. For example, a mobile phone's user interface does not require the same performance (in terms of processing speed) as its signal-processing unit. Thus the implementation strategies for these two parts are different—a software implementation is usually adequate for the user interface, whereas special-purpose hardware, in addition to a digital signal processor, may be required for the signal-processing unit. Given the trade-offs between quality of results and design level of abstraction, such performance requirements may only be met with behavioral or RTL synthesis. Models described at these lower levels need to be deployed in the context of the entire system, whose other components may be described at higher levels, even as software. In addition, accuracy of performance estimation may necessitate the use of RTL. This is particularly true of intellectual-property (IP) components supplied by third-party vendors. For example, an RTL cycle-accurate microcontroller model may be supplied in the form of a compiled library (compiled from SystemC, of course), to which we link our own designs to construct our system. It is therefore imperative for SystemC to support design methodologies based on the behavioral and register-transfer levels.

In this chapter, we will consider how to use SystemC effectively for RTL and behavioral modeling. We will see how the SystemC `SC_METHOD` and `SC_THREAD` constructs support these methodologies. Readers familiar with VHDL or Verilog may find similarities in coding styles. However, there are a few significant differences between SystemC and these languages that need careful attention. For example, using `break`s to separate `case`s in a `switch` statement may seem unwieldy to the HDL programmer, but omitting them can give rise to unintended behavior. We will also describe the hardware-oriented data-types that SystemC provides, such as arbitrary-precision integers and fixed-point types. Although some of these data-types have overlapping areas of functionality, they have different limitations and performance characteristics and hence allow the designer to select the ones that best suit each context.

4.2 Register-Transfer Level Modeling

Register-transfer level refers to that level of abstraction at which a circuit is described as synchronous transfers between functional units, such as multipliers and arithmetic-logic units, and register files. The functional units, the register files, and the connections between them are together known as the data-path. Transfers in the data-path are coordinated by a controller circuit and are synchronized to a clock. The controller's behavior is in part determined by status signals from the data-path and in part by external inputs [9]. While the structure consisting of a data-path with a controller is not unique to RTL, what distinguishes RTL from higher levels is its clock-cycle accuracy and its detailed description of combinational logic and register transfers. In other words, it is generally possible to determine, from an RTL description, the clock cycle at which each operation takes place. Sometimes, however, parts of the data-path may not be explicitly separated from the controller. For instance, arithmetic or bitwise logical expressions are commonly embedded in the description of the controller. Another characteristic of RTL is the omission of details pertaining to propagation delays of computations and data transfers. Computations and data transfers are considered to take place in "zero time," regulated by the clock. (Most designs are based on a single-phase clock, but some advanced designs may use multiple phases of a clock or multiple clocks.) Where this assumption is not valid, e.g., in the case of multicycle operations, the designer will need to allocate extra clock cycles explicitly in the design for the computation to complete. In any case, timing details within clock cycles are ignored.

4.2.1 Signals and Delta-Cycle

Let us begin by reviewing signals and the delta-cycle semantics of SystemC. Recall that a signal, being a primitive channel, supports the request–update method of access. This means that writes to the signal do not take effect immediately, but after an infinitesimal amount of time (i.e., a delta-cycle). Also, a *read* operation that is simultaneous with a *write* operation always fetches the current value of the signal, not the value that will be written to it by the *write*. Thus, it is natural to use signals to represent hardware registers in an RTL model—with the *read* operation corresponding to a connection to the output of the register and the *write* operation a connection to its input. It is important to bear in mind that a signal is not the same as a hardware register, but only *models* one when used in a particular way. When used simply to connect a combinational network with another, a signal may just model wires explicitly. Conversely, registers need not be modeled by a signal. Each synthesis tool has its rules for inference of registers, and the parts of the code that match those rules will cause registers to be generated. In general, communications between RTL processes, which

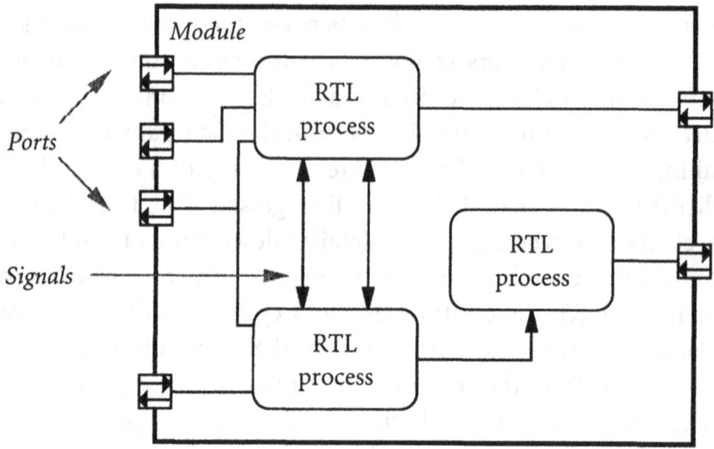

Figure 4.1: General structure of an RTL module. A module contains one or more RTL processes. Within the module, these RTL processes communicate with one another via signals internal to the module. Communication with the environment takes place through ports.

compute output and next-state functions and specify register transfers, are carried out through signals, whereas information pertinent to only one process may be stored in variables.

4.2.2 General Structure

The basic building block in SystemC is the module. A SystemC module is a container in which processes, internal channels, and other modules are instantiated. An RTL description is characterized by the kind of channels, ports, and processes it uses. In particular, as RTL is concerned with cycle-by-cycle behavior, the only channels it uses are plain and resolved signals, which do not exhibit complex temporal behavior. As a corollary, an RTL model uses only the port types `sc_in`, `sc_out`, and `sc_inout`, which correspond to `sc_signal`'s interfaces, and the resolved port types. Also, each RTL process is captured by an `SC_METHOD`. Figure 4.1 depicts the general structure of an RTL model.

4.2.3 RTL Example: A Robot Controller

We illustrate the construction of a robot controller that was deployed in a robot that solves the *shuttle puzzle*. In this puzzle, there are nine pegs and eight objects on the pegs. Four of these objects are black, the other four red. The classical problem begins with the red objects on the four leftmost and the black objects on the four rightmost pegs, with the empty peg in the middle. The goal is to move the objects in order to transpose them, according to the following rules: (1) Red objects may move only to the right, and black objects only to the left; (2) An object may move either by one step when the empty peg is adjacent to it, or by a jump over an object of the opposite color to the adjacent empty peg. This robot not only solves the classical problem, but also solves, or declares "no solution" to, problems with any initial configuration.

Description of the Controller

The robot has two degrees of freedom of movement, consisting of a horizontal conveyer belt and a vertical pulley. The conveyer belt moves the base of the puzzle back and forth, and the pulley moves an electromagnet that picks up and drops objects. Stepper motors effect movements of the conveyer belt and the pulley, and in each dimension there is a microswitch providing an absolute reference point.

The robot controller itself does not implement the algorithm to solve the puzzle. Its task is simply to generate the bit patterns that drive the stepper motors and the on/off signal for the electromagnet according to the instructions it receives from a custom-built, ROM-based microsequencer. The microsequencer, upon receiving two peg-numbers (e.g., from peg 3 to peg 5) from the algorithmic unit that solves the puzzle, computes the number of steps and direction by which the conveyer belt should move, and generates instructions for the robot controller accordingly. Each instruction consists of four bits, and the encoding is summarized in table 4.1. This layered approach allows each subsystem to be designed independently and to be potentially reused.

The interaction between the robot controller and the microsequencer is mediated through three sets of registers: a 4-bit instruction register, an 8-bit counter, and a 1-bit handshaking register. The counter is used to hold the number of steps by which the stepper motors are to move. As the counter starts counting down, the least significant bit (LSB) toggles every clock cycle, providing the pulses for the stepper motors. (The actual waveforms for the stepper motors are actually more complex, but for the present discussion this bit suffices.) When the counter reaches zero, the desired number of steps will have been produced. The inputs to the counter and to the instruction register are connected together to the bus

REPOS	MAGNET	XY	REVERSE	Action
0	0	0	0	Step X motor forwards
0	0	0	1	Step X motor backwards
0	0	1	0	Step Y motor forwards
0	0	1	1	Step Y motor backwards
0	1	–	–	Turn on electromagnet
1	1	–	–	Turn off electromagnet
1	0	0	–	Recalibrate X motor
1	0	1	–	Recalibrate Y motor

Table 4.1: Instruction encoding of the motor controller.

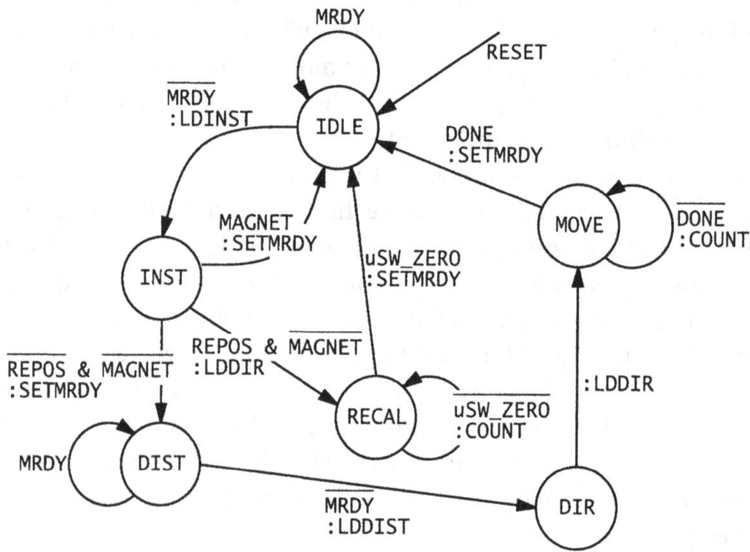

Figure 4.2: Finite state machine for the robot controller. A transition from one state to another is labeled by the condition that must hold for the transition to take place. A signal name following a colon is an output to be asserted during the transition; by default, output signals are not asserted. For example, for the transition from DIST to DIR to occur, MRDY must be zero, and if this transition does occur, the signal LDDIST will be asserted. The transition from DIR to MOVE is unconditional. Regardless of the present state, if RESET is received, the FSM enters the IDLE state.

of the microsequencer (see uSEQ_BUS in figure 4.3). Thus the loading of the instruction and the loading of the counter occur sequentially. On the other hand, not all instructions require the counter to be loaded. The instructions that recalibrate the motors and those that turn on or off the electromagnet do not need it at all. Finally, the microsequencer announces the availability of the instruction or the step count on the bus by clearing the handshake register MRDY. The finite state machine shown in figure 4.2 describes how the robot controller interacts with the microsequencer and how instructions are translated into actions.

Upon reset, the FSM enters the IDLE state, and waits till the MRDY register is cleared. It then loads the instruction word from the microsequencer by asserting the LDINST (*load instruction*) signal and enters the INST state. Then, depending on the instruction bits, the FSM does one of three things:

(1) If the instruction pertains to the electromagnet, the FSM simply sets the MRDY register and returns to the IDLE state, since another circuit actuates the control of the magnet.

(2) If the instruction pertains to recalibration, the FSM enters the RECAL state, and then continually asserts the COUNT signal, causing the counter to count down, until the microswitch for the dimension selected by XY is pressed (as indicated by the signal uSW_ZERO).

(3) If the instruction is to drive either motor by a designated number of steps, the FSM will go through a sequence of states before enabling the counter to count down. First, in state DIST it loads the number of steps into the counter by asserting LDDIST (*load distance*). It then sets the direction of the movement in state DIR by asserting LDDIR (*load direction*). Finally, it stays in the MOVE state until the counter reaches zero (as indicated by the signal DONE). Thus, when the conveyor belt or the vertical pulley reaches its destination, the FSM returns to the IDLE state and sets MRDY to indicate that the controller is now ready for another instruction.

SystemC Model of the Controller

Having examined the functionality of the robot controller, let us now proceed to write an RTL description for it. The first question we ask is: What are the inputs and outputs to the module that we are about to describe? First, as is common in many RTL modules containing FSMs, we need a clock and a reset. Then we need a port connected to the microsequencer's bus and another for microsequencer to clear the handshake signal MRDY, as well as the signal uSW_ZERO that is derived from the microswitches. For outputs, we need to provide the following to the circuitry for the motors and the magnet: (1) the instruction word held in the instruction register; (2) the LSB of the counter, which provides the pulses for the stepper motors; and (3) the signal LDDIR, which sets the direction of the movement. We also need to make the handshake signal MRDY visible to the microsequencer. Now since all of these outputs are also used within the controller module itself, we declare them as in/out ports instead of output ports. The declaration of the ports is shown below, the commented sections to be expanded shortly:

```
SC_MODULE(robot_controller) {
    sc_in<bool>          CLOCK;
    sc_in<bool>          RESET;
    sc_in<sc_bv<8> >     uSEQ_BUS;
    sc_in<bool>          CLRMRDY;
    sc_in<bool>          uSW_ZERO;

    sc_inout<bool>       MRDY;
    sc_inout<bool>       REPOS;
    sc_inout<bool>       MAGNET;
    sc_out<bool>         XY;
    sc_out<bool>         REVERSE;
    sc_out<bool>         LDDIR;
    sc_out<bool>         LSB_CNTR;

    /* Internal variables and signals */
    /* Member function prototypes */
    /* Constructor */
};
```

How do we treat the other output signals of the FSM that we have just described, such as LDINST and LDDIST? Since the instruction register and the counter are part of the module, and these signals are not used elsewhere, we declare them not as ports, but as internal signals. Furthermore, since we are describing the FSM explicitly, we need two internal signals to hold the current state and the next state. Here signals are required because the FSM is a *Mealy machine*, whose outputs depend not only on the current state but also directly on the input. Changes in the current state are to be propagated to the output in the same cycle as changes in the input, and signals (along with sensitivity) ensure that the propagations take place.

To represent the state, we use an enumerated type that mirrors the symbolic names such as IDLE and INST that we have used in figure 4.2. Then we declare the internal variables and signals as follows:

```
/* Internal variables and signals */
```

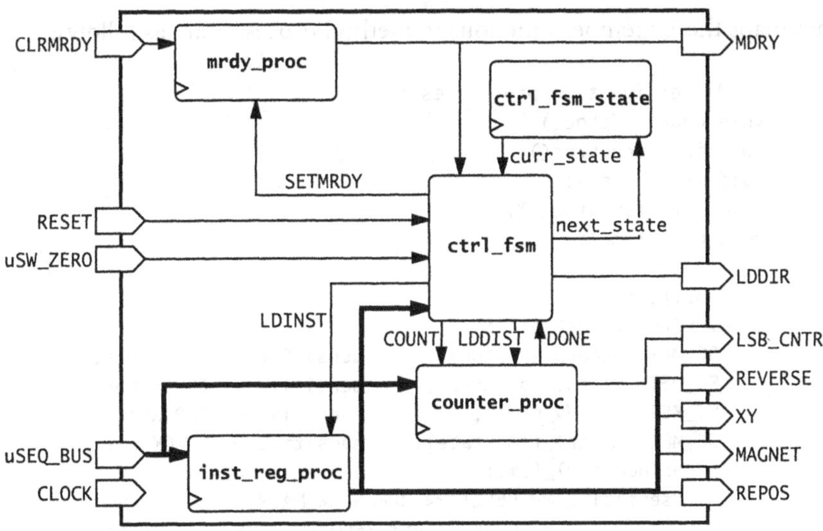

Figure 4.3: Structural diagram of the robot controller. Each block represents an RTL process within the controller module. A connection between two blocks, if not also connected to a port, requires an internal signal (e.g., COUNT and DONE). For clarity, connections to the clock input are not explicitly drawn; the small angle bracket on the lower-left corner of a block denotes sensitivity to the positive edge of the clock.

```
enum ctrl_state { IDLE, INST, DIST, RECAL, DIR, MOVE };
sc_signal<ctrl_state> curr_state, next_state;
sc_signal<bool> DONE, LDDIST, COUNT;
sc_uint<8> counter;
sc_signal<bool> LDINST, SETMRDY;
```

Observe that, in contrast to the other members, counter is declared as a variable, not as a signal. This is because it is used exclusively within one of the processes of the module. We will elaborate on this point in section 4.2.4.

Next, we need to determine what processes are necessary. In the foregoing discussion, we saw that the robot controller is responsible for maintaining a handshake register, an instruction register, and a counter. Each of these registers is updated upon the rising edge of the clock; hence, it is natural to allocate one process for each. We also need a process for computing the output and next-state functions, as well as one for updating the next state. Figure 4.3 shows the structure of this division of responsibilities, and the corresponding prototypes of the member functions and a constructor that maps these member functions to method processes are as follows:

```
/* Member function prototypes */
void counter_proc();
void inst_reg_proc();
void mrdy_proc();
void ctrl_fsm_state();
void ctrl_fsm();

/* Constructor */
SC_CTOR(robot_controller) {
    SC_METHOD(counter_proc);       sensitive << CLOCK.pos();
    SC_METHOD(inst_reg_proc);      sensitive << CLOCK.pos();
    SC_METHOD(mrdy_proc);          sensitive << CLOCK.pos();
    SC_METHOD(ctrl_fsm_state);     sensitive << CLOCK.pos();
    SC_METHOD(ctrl_fsm);
        sensitive << RESET << REPOS << MAGNET
                  << DONE << uSW_ZERO << MRDY << curr_state;
}
```

We could have also included the body of the member functions inside the SC_MODULE definition, but it is customary to define member functions separately. Let us first consider counter_proc(), which describes the operation of the counter and produces the DONE and LSB_CNTR signals.

```
void robot_controller::counter_proc()
{
    if (LDDIST.read()) {
        counter = uSEQ_BUS.read();
    } else if (COUNT.read()) {
        counter = counter - 1;
    }
    DONE.write(counter == 0);
    LSB_CNTR.write(counter[0]);
}
```

Upon the rising edge of the clock, if the signal LDDIST is asserted, the value on the microsequencer's bus is loaded into the counter. Otherwise, if the signal COUNT is asserted, the counter's value is reduced by one. In addition, the signal DONE receives the value one if and only if the counter is zero, and LSB_CNTR (which, we recall, is an output that provides the pulses for the stepper motors) simply mirrors the LSB of the counter. Similarly, the process inst_reg_proc() loads into a temporary variable inst a 4-bit instruction from the bus if LDINST is asserted. The instruction is then deconstructed and copied to the output ports REPOS, MAGNET, XY, and REVERSE.

```
void robot_controller::inst_reg_proc()
{
    if (LDINST.read()) {
        sc_bv<4> inst = uSEQ_BUS.read();
        REPOS.write(   inst[0].to_bool());
        MAGNET.write(  inst[1].to_bool());
        XY.write(      inst[2].to_bool());
        REVERSE.write( inst[3].to_bool());
    }
}
```

The next two processes are simpler: mrdy_proc() sets the value of MRDY according to CLRMRDY and SETMRDY, and ctrl_fsm_state() just updates the value of curr_state.

```
void robot_controller::mrdy_proc()
{
    if (CLRMRDY.read()) {
        MRDY.write(false);
    } else if (SETMRDY.read()) {
        MRDY.write(true);
    }
}

void robot_controller::ctrl_fsm_state()
{
    curr_state.write(next_state.read());
}
```

Finally, we come to the definition of the process that computes the output and next-state functions: ctrl_fsm(). Unlike the previous four processes, ctrl_fsm() is not triggered on the rising edge of the clock. Instead, the functions that it computes are purely combinational, dependent directly on its inputs. This is reflected in the sensitivity declaration in the constructor. The SystemC description closely follows the graphical one in figure 4.2.

```
void
robot_controller::ctrl_fsm()
{
    control_state ns = curr_state;

    LDDIST.write(false);
    COUNT.write(false);
    LDINST.write(false);
    SETMRDY.write(false);
```

```
if (RESET.read()) {
    ns = IDLE;
} else {
    switch (curr_state.read()) {
    case IDLE:
        if (! MRDY.read()) {
            LDINST.write(true);
            ns = INST;
        }
        break;
    case INST:
        if (MAGNET.read()) {
            SETMRDY.write(true);
            ns = IDLE;
        } else if (REPOS.read()) {
            LDDIR.write(true);
            ns = RECAL;
        } else {
            SETMRDY.write(true);
            ns = DIST;
        }
        break;
    case RECAL:
        if (uSW_ZERO.read()) {
            SETMRDY.write(true);
            ns = IDLE;
        } else {
            COUNT.write(true);
        }
        break;
    case DIST:
        if (! MRDY.read()) {
            LDDIST.write(true);
            ns = DIR;
        }
        break;
    case DIR:
        LDDIR.write(true);
        ns = MOVE;
        break;
    case MOVE:
        if (DONE.read()) {
            SETMRDY.write(true);
            ns = IDLE;
        } else {
            COUNT.write(true);
        }
        break;
    } /* switch */
```

```
    }
    next_state.write(ns);
}
```

Several things are noteworthy here. First, at the beginning we unconditionally assign a *default value* to the output signals. Recall in the description of the FSM in section 4.2.3 (figure 4.2) we specified that if an output signal is not shown on a transition, it should receive the value *false*. While we could specify this informally in the graphical description, we need to state this explicitly in the SystemC model, because the semantics of signals are such that it retains its value until a process changes it. One way to ensure this is to make assignments to all output variables in all branches of execution. This is, however, very cumbersome and error-prone. Instead, it is better to use default values and then override them where required. There are two ways to do this. The first is to use a temporary variable (initialized with the default value), and then transfer its value to the corresponding output at the end; in the preceding example, the temporary variable `ns` is utilized on behalf of the output `next_state`. The other is to operate on the signals directly, as we have done for `LDDIST`, `COUNT`, etc.

Next, when we read the value from a signal, we use the method `read()`. In many cases `read()` may be omitted, but not always. If the underlying data-type of the signal needs a non-built-in conversion before being used in the context, then `read()` is required because C++ does not allow two consecutive implicit non-built-in conversions (§12.3 of the ISO/IEC 14882 C++ Standard [13]). For instance, the following would cause a compile-time error because it requires a conversion from a signal type to its underlying data-type, and then from `sc_bv<4>` to `sc_uint<4>`:

```
    counter = uSEQ_BUS;
```

While the use of `read()` appears painful, it helps to underline the fact that the object in question is a signal and not a variable, and avoids problems with conversions. On the other hand, once the designer is comfortable with these arcane rules of C++, he may choose to omit it if he considers its use too verbose.

Finally, at the top level of `ctrl_fsm()` we have an `if`-statement based on `RESET`. The true branch sets the next state variable `ns` to the `IDLE` state, as expected. The `else` branch consists of a `switch` statement on `curr_state`, and each `case` of the `switch` computes the next-state and output functions based on the primary inputs and the current state. This form is extremely common in descriptions of FSMs with reset, and corresponds exactly to the graphical representation of figure 4.2.

4.2.4 Traps and Pitfalls

One of the most annoying pitfalls of SystemC for those versed in Verilog or VHDL is that, in SystemC, a branch labeled by `case` *falls through* to the next branch unless a `break` statement terminates it. In our robot controller example, therefore, we have been careful to place a break statement at the end of each of the six cases. Even if a fall-through would implement the correct functionality, such a construct ought to be avoided in general. This makes the branches mutually independent, and modification of one branch does not unintentionally affect another.

Another common mistake is to expect to obtain the new value of a signal or a port after an assignment has been made to it. Like VHDL's signal assignment and Verilog's nonblocking assignment, SystemC's signal assignment does not take effect immediately. Thus, for example, if we were to read `LDINST` in state `IDLE` after the statement `LDINST.write(true)`, we would not obtain the value just written, but the value `LDINST` had previously. If an output value is to be used internally as well, we should use temporary variables (e.g., `ns`) in all operations and transfer their values to the corresponding outputs. (See section 2.10.)

Also, the difference between a signal and a variable must be heeded. If an object is used exclusively in one process, a plain member variable of the enclosing module suffices. For instance, `counter` is used only in the process `counter_proc()` and is therefore declared as a plain member variable. However, if the value is to be shared among two or more processes, e.g., `next_state`, then a signal is necessary. This is because the order in which the processes are triggered is undetermined—if one process reads a variable and another process writes to it, we cannot know whether the write occurs before or after the read, and this makes a great difference. A signal, thanks to the request–update semantics, is not affected by the order of execution.

With respect to synthesis, if a process both reads from and writes to a signal, or if it assigns to a signal in some but not all possible paths of execution, latches or flip-flops may be produced as a result. Thus a designer must take heed that purely combinational circuits such as `ctrl_fsm()` do not have these constructs. A good synthesis tool will warn the designer of this possibly undesired consequence.

Purely combinational processes should also be made sensitive to all of their inputs; otherwise, incorrect simulation may result because changes in some inputs may not be propagated. Conversely, redundant sensitivities ought to be avoided so that computations do not take place unnecessarily.

This is an area where a good synthesis tool or syntax checker will be helpful also.

Lastly, member variables (such as `counter`) and signals (such as `DONE`, `LDDIST`), if they are not initialized in the constructor of the module, will have indeterminate values in the beginning. Unless they are of multivalued logic types (e.g., `sc_logic` and `sc_lv<W>`, section 4.4.4), in which there is a distinct `'X'` to denote uninitialized value, actual values may vary from one compilation to another. This is usually innocuous if reset signals are appropriately employed; but if uninitialized values are monitored and written to a log, differences may arise in the first few cycles of simulation. Yet it is better to initialize member variables and signals, along with output and in/out ports, in the constructor. If we need to verify that the hardware is robust against randomness of initial values, we may simulate with multiple configurations, each having different yet deterministic initial values.

4.3 Behavioral-Level Modeling

Over the years the phrase *behavioral level* has taken on various meanings. At the time when RTL modeling and synthesis were rising to prominence, behavioral level has sometimes been used synonymously with RTL. That usage, however, was to contrast RTL with lower levels such as schematic capture. As higher-level design methodologies began to emerge, a more precise meaning of behavioral level has been adopted—namely, the level at which the primary concern is the relative ordering of input and output events rather than exact clock cycles.

Thus, the most significant distinction between behavioral level and RTL is that in RTL the designer must decide what the circuit states are, how the circuit transitions from one state to another, and what operations take place in each state. In addition, in RTL the designer must determine in advance how many resources (functional units and registers) to use and how to share them. This greatly limits the flexibility of design exploration—RTL designs are difficult to modify because a small change in one place may necessitate changes in many others.

In contrast, behavioral-level modeling raises the abstraction level by allowing us to think of our design as program flow. External behavior is defined by the sequencing of input and output events, not by clock cycles. The mappings of internal operations to clock cycles (scheduling) and functional units (resource binding) are not relevant here—these "details" are later worked out either manually or automatically with a synthesis tool according to user-specified constraints. Behavioral level, however, shares one common trait with register-transfer level: the pins at structural boundaries are delineated. Hence, we sometimes refer to them as *pin-accurate levels* from the structural perspective.

4.3.1 Basics

In this chapter we will consider only *synchronous* behavioral modeling, in which input and output events are sequenced and synchronized by a clock and signals are the primary communication channels. (More abstract modeling methodologies, such as transaction-based modeling, will be discussed in chapter 8 and chapter 9.) The general structure of synchronous behavioral models is similar to that shown in figure 4.1.

Even though the use of the clock appears to be a reversion to RTL, it is not. The clock is employed only as an ordering and synchronizing mechanism, and a clock cycle for a behavioral model may be mapped to one or more actual clock cycles in the final implementation. In SystemC, we use SC_THREAD to map a member function to a behavioral process. To make it synchronous, we specify that the thread process be sensitive to either the positive edge or negative edge of a clock; but the process should not be made sensitive to any other events.

For the purpose of this section, we introduce another construct called *clocked thread process* (SC_CTHREAD), which is in many ways similar to the thread process (SC_THREAD), except that its sensitivity is restricted to a single edge of a clock, and that it provides two useful constructs: `watching()` and `wait_until()`. The `watching()` construct allows for (among other things) concise modeling of the synchronous reset (see the example in section 4.3.2), and `wait_until()` facilitates handshaking. (SC_CTHREAD was introduced in an earlier version of SystemC. At the time of this writing, the OSCI Language Working Group has yet to formulate a more general mechanism for interrupt handling. When that work comes to fruition, SC_CTHREAD, along with `watching()` and `wait_until()`, will eventually be subsumed by SC_THREAD. Note also that SC_CTHREAD processes are not executed during the initialization step of simulation [step 1 of section 2.10].)

A behavioral process separates its I/O cycles by either a `wait()` or a `wait_until()` statement. We define an *I/O cycle* as an interval of time in which there is no exchange of data between the process and the environment, either sampling of inputs or production of outputs; exchanges of data occur on I/O cycle boundaries. A `wait()` (without parameters) causes the process to suspend itself, to be resumed later when the next clock comes along. A `wait_until()`, on the other hand, allows the designer to specify a condition (though somewhat limited in expressiveness) that must take place for the process to be resumed. The condition is synchronized to the clock—the values of the signals in the condition expression are sampled only on clock edges, and thus the reaction of the process to the changes in signal value is not immediate, but deferred to the upcoming clock edge.

4.3.2 Behavioral Example: Euclid's Algorithm

Let us consider an example that computes the greatest common divisor (GCD) of two nonnegative integers (but not both zero), using Euclid's algorithm. Euclid's algorithm may be stated as follows: Given two integers a and b, where $a \geq 0$ and $b > 0$,

- If b divides a, then $\text{GCD}(a, b) = b$;
- Otherwise, $\text{GCD}(a, b) = \text{GCD}(b, r)$ where r is the remainder of a divided by b.

Our module has a structural boundary consisting of ports for the clock, the reset, the two inputs A and B, the output C, and a handshake signal READY. For this exposition we first use a simplified protocol to communicate with the environments by asserting READY in the same cycle when the output is valid on port C, and then proceeding to sample inputs in the next cycle and lowering READY. A real design should use a more robust handshake protocol so that the timing of the sampling of inputs can be controlled as well, as shown in section 4.3.3. Here is the definition of the `euclid_gcd` module:

```
SC_MODULE(euclid_gcd) {
    sc_in_clk         CLOCK;
    sc_in<bool>       RESET;
    sc_in<unsigned>   A, B;
    sc_out<unsigned>  C;
    sc_out<bool>      READY;

    void compute();
    SC_CTOR(euclid_gcd) {
        SC_CTHREAD(compute, CLOCK.pos());
        watching(RESET.delayed() == true);
    }
};
```

We use the SystemC construct SC_CTHREAD to map the member function `compute()` to a clocked thread process. Recall that a thread process allows implicit state keeping. That is, we may use `wait()` statements to suspend the execution of the process until an event occurs to which the process is sensitive. Thread-process descriptions tend to reflect more naturally the overall flow of the algorithm to be implemented.

In the SC_CTHREAD declaration we have specified that this process is sensitive to the positive edge of the clock. The next declaration, `watching()`, specifies that this process should be restarted whenever the input RESET is asserted. This is a *synchronous reset*, meaning that the process is restarted on the next rising clock, not immediately after the assertion of RESET. The definition of `compute()` is as follows:

```
void
euclid_gcd::compute()
{
    // reset section
    unsigned tmp_a = 0, tmp_b;

    while (true) {           // main loop

        // Here is an I/O cycle
        C.write(tmp_a);      // place output on C
        READY.write(true);   // and assert READY
        wait();

        // Another I/O cycle starts here
        tmp_a = A.read();    // sample inputs
        tmp_b = B.read();
        READY.write(false); // lower READY
        wait();

        // No I/O takes place during the computation of GCD.
        // Computation and communication are separated
        while (tmp_b != 0) {         // Euclid's algorithm
            unsigned r = tmp_a;
            tmp_a = tmp_b;
            r = r % tmp_b;           // compute remainder
            tmp_b = r;
        }
    }
}
```

The heart of the process `compute()`, the loop labeled *Euclid's algorithm*, is written just as it would be in a conventional software programming environment. No special SystemC construct was used here. To compute the remainder, we have simply used the native % operator. If synthesis is the goal of this design and the synthesis environment does not support the native % operator, we will have to refine our model further by replacing it with successive subtraction. This results in a model that executes more slowly during simulation, because more implementation details are taken into account:

```
while (r >= tmp_b) {          // compute remainder
    r = r - tmp_b;
}
```

The rest of the process pertains to the communication with the environment: moving data in and out and signaling the validity of the output. After assigning zero to tmp_a upon reset, the process enters the *main loop*, which iterates indefinitely until another reset comes along. It then transfers the value in tmp_a to the output port C, at the same time raising READY. In the next I/O cycle, new inputs are sampled from the input ports A and B, and READY is lowered. (We have "wrapped around" the outputting of the result to the beginning of the main loop; we could have placed it after the *Euclid's algorithm* loop and duplicated it in the *reset section.*)

Note especially the use of wait() to separate I/O cycles. If, for example, the first wait() had been omitted, the first assignment to READY would have been superseded by the second, and it would have been impossible to tell when the output was ready and when the input was sampled. Note as well that there are no wait() statements in the *Euclid's algorithm* loop. Since this loop does not perform any input or output operations (read and write), adding waits does not change the behavior of this module insofar as this abstraction level is concerned, even though simulation will take longer. (Recall that only the relative ordering of I/O events is relevant to the behavioral level.)

Contrast this example with the robot controller example in section 4.2.3. At the behavioral level, our view of the design focuses on the overall program flow. We do not have explicit state variables (such as curr_state) to indicate where we are in the flow of computation. For instance, for the two operators in *Euclid's algorithm* of the refined version, >= and -, we have no information regarding the exact clock cycles in which they take place in the final implementation. In general, if a synthesis tool is used to generate an implementation from a behavioral description, different user-specified constraints will result in different cycle-by-cycle behaviors, with certain trade-offs (e.g., between area and cycle count).

At RTL, however, our focus would instead be on circuit states, how one state leads to another, and what control signals the circuit should produce. If the robot controller had more arithmetic computations, we would not see an algorithm described naturally as program flow; rather, we would have a predesigned data-path and signals emanating from the finite state machine to control operations and data transfers therein.

4.3.3 Traps and Pitfalls

The `wait()` and `wait_until()` statements play a central role in constructing synchronous behavioral models. It is important that signal or port assignments belonging to different I/O cycles be separated by one or more `wait()` or `wait_until()`. Otherwise, the earlier assignments would be superseded by later ones, and the net effect would be that of the latter.

Since input/output events define the behavior of a behavioral process and exact cycles are unknown, we recommend that handshaking schemes be used along with the exchange of data. Handshaking ensures that correct interaction with the environment is not affected by the "stretching" of I/O cycles in absolute time. In the example of section 4.3.2, we used a READY signal to inform the caller of the module of the availability of the output, and, assuming that the caller would be in a state to grab the result and provide the next inputs, we go on to sample new inputs in the next cycle. This assumption may be inadequate. A more robust handshake protocol is to have the caller indicate the availability of new inputs through another signal, say START. Then, instead of a simple `wait()`, we would state:

```
C.write(tmp_a);              // place output on C
READY.write(true);           // and assert READY
wait_until(START.delayed() == true);
```

The caller, for its part, will wait for `euclid_gcd`'s READY signal to be raised before it asserts START and places new input values on A and B.

The construct `wait_until()`, like `watching()`, accepts an expression consisting of signals, ports, variables, and constants, all of which must be of boolean type. We note that when a signal or port is used in such expressions, the method `delayed()` should be called, so that the signal or port may be resampled at every clock. If `delayed()` is omitted, the value of the signal is sampled in the current cycle and the same is used throughout the duration of the `wait_until()`.

Another implication of inexact cycle behavior is that correctness cannot be defined as cycle-by-cycle match in simulation in general. If a behavioral model is manually refined to RTL or automatically synthesized, the resulting model may have I/O cycles mapped to multiple clock cycles. For instance, any reasonable implementation of the *Euclid's algorithm* loop is likely to require multiple (even a data-dependent number of) cycles to compute, even though at the behavioral level it spans only one I/O cycle. Consequently, we ought to write our testbenches with the same handshaking principles, so that we may reuse them for both the original model and the refined or synthesized one.

Infinite loops must have at least a `wait()` or `wait_until()` statement in every path through the loop. Otherwise, the thread process will monopolize the whole simulation. This also makes great sense with respect to I/O cycles—for what does it mean to have an infinite loop that does not possibly affect any I/O?

Exceptions, as modeled by `watching()`, present situations that may require additional `wait()` statements. In the `euclid_gcd` example, we used `watching()` to model top-level reset. But control flow may be diverted back to the beginning of the thread process `compute()` only at I/O cycle boundaries. For instance, when the thread process enters the Euclid's algorithm loop, it is not possible for it to react to the reset signal. If we wish to test the reset behavior at a finer granularity, we need to insert `wait()` statements accordingly.

Finally, as we have noted in chapter 2, the use of `wait()` statements in `SC_THREAD` and `SC_CTHREAD` processes incurs some performance penalty due to context-switching activities. Therefore, while it is possible (and sometimes even necessary) to write cycle-accurate descriptions with thread processes or clocked thread processes, whence a `wait()` is required for every clock cycle instead of just every I/O cycle, the potential impact on simulation performance needs to be borne in mind.

4.4 Hardware-Oriented Data-Types

The standard built-in C++ data-types (also called *native types*) lack the flexibility needed to model hardware designs. They come only in a small set of predetermined sizes—on most platforms, usually powers of two. In hardware, however, one is often compelled by area and performance objectives to use "nonstandard" widths. Moreover, in certain situations data-types of widths larger than the native set could represent may be necessary. Sometimes software implementations may also require nonnative data-types. For example, fixed-point types are widely employed in DSP algorithms, which might be mapped to software executing on a DSP core or to DSP hardware with direct support for fixed point. To this end, SystemC provides a rich collection of data-types, along with overloaded standard arithmetic operators that make the expression syntax similar to that of native data-types. In this section we will present an overview of the SystemC data-types, with an emphasis on the usage and cautionary notes. The reader should consult the *SystemC User's Guide* [36] for complete details.

4.4.1 Fixed-Precision Integral Types and Operations

Let us first consider fixed-precision integral types. In conventional software programming, integral variables are usually of `int` or `unsigned` type. This is true even if the range of the type is larger than necessary. For instance, a loop index may range from zero to 31, whence six bits would be sufficient (one extra bit for it to hold the value 32 to indicate end-of-loop), but it is still customarily held in an `int`. The reason is that there is nothing to be gained in performance and little in data size if we use a narrower data-type instead. In hardware, however, redundant bits may result in redundant registers and operators that are larger than necessary. For the loop index, we needed only a 6-bit register and a 5-bit adder; a 32-bit register and a 32-bit adder would be wasteful. Even if an automatic tool is capable of optimizing variable and operator widths under certain conditions, there are always circumstances where only the designer knows the dynamic range of the variables in his design.

SystemC provides a set of fixed-precision integral types, `sc_int<W>` and `sc_uint<W>`. These types are parameterizable by the user, with `W` denoting the width of the type in bits. Thus, for example, `sc_int<4>` is a 4-bit, signed type (ranging from -8 to 7), and `sc_uint<6>` is a 6-bit unsigned type (ranging from 0 to 63).

In the SystemC reference implementation, these types are based on the extended integral types. While these extended types (usually called `long long` and `unsigned long long`) do not belong to the C++ standard, most compilers support them as part of the collection of built-in types. On a 32-bit machine, these types are 64 bits wide. As a result, operations involving `sc_int<W>` and `sc_uint<W>` are limited to 64 bits. That is to say, intermediate results that exceed 64 bits will be truncated so that only the least significant 64 bits remain. Likewise, on assignment from a wider type to a narrower type, the significant bits are lost, including the sign bit if the source operand is of a signed type.

With respect to simulation performance, `sc_int<W>` and `sc_uint<W>` are the fastest types in SystemC, second only to the native types. Therefore, we recommend that the designer use these types (and of course the native types) as much as possible.

4.4.2 Arbitrary-Precision Integral Types and Operations

In some situations 64 bits may not be sufficient, and therefore we need data-types in addition to `sc_int<W>` and `sc_uint<W>`. In still other occasions we might require intermediate results that are larger than 64 bits, even if end results could fit in fewer. SystemC provides two arbitrary-precision integral types, `sc_bigint<W>` and `sc_biguint<W>`. They behave in much the same way as their fixed-precision counterparts, but intermediate results may have arbitrary precision. For instance, consider the following expression:

```
sc_uint<40> x, y, z, w;
w = (x * y) / z;
```

Since intermediate results are limited to 64 bits for `sc_uint<W>`, `(x * y)` may become truncated before being divided by `z`, and thus the value assigned to `w` may not be what the designer expects. To alleviate this problem, we need either to declare these variables as `sc_biguint<40>`, or to use a temporary variable of type `sc_biguint<80>`:

```
sc_biguint<80> t = x;
t *= y;
w = t / z;
```

Note that the assignment `t = x * y;` would not produce the desired result even though `t` has sufficient precision, because the precision of `x * y` would have already been limited to 64 bits. Another subtlety is introduced by the subtraction in `sc_biguint<W>`. If `x` and `y` are of type `sc_biguint<20>`, and `x` is 1 and `y` is 2, what is the value of `(x - y)`? Strange as it might seem, it is "-1". As an intermediate value not yet assigned to a variable, `(x - y)` retains "all the bits" that are required so that, when finally assigned to a variable, no precision is lost. We may conceive of a "negative" `sc_biguint` as a 2's complement number consisting of an infinite number of 1s to its left. Hence, if we assign `(x - y)` to variables:

```
sc_biguint<30> u = x - y;
sc_biguint<40> v = x - y;
```

we will have $\mathsf{u} = 2^{30} - 1$ and $\mathsf{v} = 2^{40} - 1$.

The ability to keep arbitrary precision during operations comes with a price. These arbitrary precision types are slower than their fixed-precision counterparts and therefore should be used only sparingly.

4.4.3 Bits and Bit-Vectors

In our examples we have used the native type `bool` to represent single-bit values. We have also used `false` and `true` instead of `0` and `1`, but either choice is perfectly fine. An idiosyncrasy of the `bool` type may give rise to subtle problems. Even though conceptually `bool` variables may take on no other values than `0` and `1`, in many C++ environments an uninitialized bool variable may violate this assumption. As a general rule, use of uninitialized variables (especially `bool` variables) should be avoided. Also, the *logical* NOT (`!`) should not be confused with the *bitwise* NOT (`~`); `!true` is `false`, but `~true` is 4294967294 (on a 32-bit platform), which, when converted back to `bool`, remains `true`.

Besides the arithmetic types, SystemC also provides a type for bit-vector manipulation: `sc_bv<W>`. Like `sc_bigint` and `sc_biguint`, the width `W` for `sc_bv` is unlimited. But `sc_bv` does not provide arithmetic operations such as addition and multiplication; it is designed and optimized primarily for bit-oriented operations. Conversions between the arithmetic types and the nonarithmetic bit-vector type `sc_bv` are provided, and only a simple assignment is necessary to effect a conversion.

Operands to the bitwise AND (&), OR (|), and XOR (^) operators must be of the same width. The same is required of assignments. Where there are mismatches in width, concatenation and part-selection may be used. Assignments from both string and integral constants are allowed; for example,

```
sc_bv<16> x = "1011011011010001";
sc_bv<12> y = 3921;     /* 111101010001 */
```

In addition, *reduction operations* are available by way of these functions: `and_reduce()`, `or_reduce()`, and `xor_reduce()`. For instance, to determine the parity (that is, the XOR of all the bits) of a bit-vector `databus`, we write:

```
sc_bv<64> databus;
bool result = databus.xor_reduce();
```

4.4.4 Four-Valued Logic and Logic-Vectors

The data-types that we have examined in the previous sections work well for modeling parts of the design where arithmetic or logical operations take place. In other words, they are *computation-oriented*. However, for parts of the design that need to be modeled with tristate capabilities, additional data-types are called for. To this end SystemC provides two data-types: `sc_logic` and `sc_lv<W>`. The former is a single-bit type and the latter a vector type. In contrast to the computation-oriented types, which are based on binary values, these types are 4-valued. What this means is that each bit may take two other possible values than `0` and `1`, namely, `X` and `Z`. `X` represents the indeterminate value, and `Z` represents the tristate high-impedance. If a 4-valued (bit or vector) variable is not initialized explicitly, its value will be `X` or a vector of `X`s.

Assignments to variables of type `sc_logic` may take character literals, in addition to expressions of type `sc_logic`, e.g.,

```
sc_logic x, y;
x = '1';
y = 'Z';
```

Also, as with `sc_bv`, assignments from string and integral constants to `sc_lv` are allowed, with the string constant consisting of the characters `'0'`, `'1'`, `'X'`, and `'Z'`. Operations for `sc_lv` are analogous to those for `sc_bv`.

Conversions from the arithmetic and bit-vector types are straightforward. One must be cautioned, however, of the possible loss of information resulting from conversions in the other direction, as the values `X` and `Z` cannot be represented in the target types. Therefore, the SystemC library will generate warning messages if there should be such loss of information. These warning messages should not be ignored. In traditional HDL modeling, `X`s are often used to track down design errors (such as reading from an uninitialized register or driving a bus with conflicting values), but since the propagation of `X`s stops at these conversions, these messages take on the role of propagation.

4.4.5 Resolved Types

Four-valued types find most of their uses in the modeling of busses. Now recall that a plain signal has at most one driver and may have more than one reader. (A driver of a signal is a SystemC process, whether `SC_METHOD` or `SC_THREAD`, that writes to the signal.) In contrast, a bus may have many drivers and many readers. Since different drivers may attempt to place different values on the bus, a resolution function is needed to determine the final value. For the 4-valued types `sc_logic` and `sc_lv`, SystemC provides the corresponding resolved signal types, `sc_signal_resolved` and sc_signal_rv<W>. There are also port types for connecting to resolved signals: sc_inout_resolved and sc_inout_rv<W>.

Resolution for the 4-valued types is intuitive: (1) If all non-Z values agree, the resulting value is that non-Z value; (2) if some non-Z values conflict (e.g., a 1 and a 0 driven at the same time), then the resolved value is X; (3) otherwise, if all processes write Z to the signal, then the resolved value is Z. Resolution works bitwise, that is, each bit of an sc_signal_rv is resolved independently of others. These rules are identical to Verilog's.

Note also that each driver retains the value that it has previously driven until explicitly changed. Thus, for example, if resolved signals or ports are used in an RTL process, it is prudent for the designer always to write a "default value" (usually Z) onto the signal or port and override it later (cf. ctrl_fsm() in section 4.2.3, on page 59).

4.4.6 Fixed-Point Types

Floating-point numbers are extremely useful for modeling arithmetic operations at a high level because they can support a large range of values and are easily scaled. Yet they are not hardware-friendly: floating-point units are large and complex, consume much power, and are notoriously difficult to verify. Thus floating-point arithmetic is typically limited to microprocessors and high-end digital signal processors. For most other applications fixed-point arithmetic is used instead.

SystemC provides a library for fixed-point arithmetic that consists of a comprehensive set of operations and conversions, as well as a variety of rounding (or *quantization*) and saturation modes. In this exposition, we will focus on the template type sc_fixed<wl, iwl, qm, om, nb>, which represents signed fixed-point numbers. The meanings of the parameters are as follows: wl, word length; iwl, integer word length; qm, quantization mode; om, overflow mode; and nb, number of bits for overflow. The last three parameters are optional, and nb is meaningful only if the om selects one of the wrap-around modes.

Word length specifies the total precision of the type and must be positive. *Integer word length* specifies the position of the binary point relative to the most significant bit of the word; it may be positive, zero, or negative. If iwl is greater than wl, then zeros are padded to the right; if iwl is negative, then the sign bits are extended to the left of the word up to the binary point. The following table shows four different configurations, where b denotes significant bits and s denotes the sign bit. Note also the position of the binary point.

Representation of value	wl	iwl
`bbbbb00.`	5	7
`.ssbbbbbb`	6	-2
`.bbbbbbbbb`	9	0
`bbb.bbbb`	7	3

Operations on fixed-point data-types are performed in arbitrary precision. When the result is assigned to a fixed-point entity, two adjustments may take place. First, quantization occurs if the precision of the target is too coarse-grained. For instance, `sc_fixed<7,3>` may represent values up to a precision of 2^{-4}, and anything finer than that must be rounded. The direction of the rounding is determined by the parameter `qm`; some of the available options are: toward $+\infty$ (`SC_RND`), toward zero (`SC_RND_ZERO`), toward $-\infty$ (`SC_RND_MIN_INF`), convergent rounding (`SC_RND_CONV`, i.e., to the nearest value), and truncation (`SC_TRN`, the default).

The other adjustment is concerned with overflow. When a result is outside of the range of the target type, it is adjusted according to the parameters `om` and `nb`. There two basic kinds of adjustments, *saturation* and *wrap-around*, each with variations. Let us denote by *min* and *max* the minimum and maximum values that may be represented by the target type.

For saturation, one may choose from three variants: *simple* (`SC_SAT`), *symmetrical* (`SC_SAT_SYM`), and *zero* (`SC_SAT_ZERO`). In simple saturation mode, values greater than *max* are reduced to *max*, and values smaller than *min* are increased to *min*. Symmetrical saturation is similar to simple saturation, except that for negative numbers $-max$ is used instead of *min*. In saturation-to-zero mode, out-of-range values just become zero.

As for wrap-around, we will note only briefly that the basic idea is to discard those most significant bits that cannot fit in the target type. Variants of wrap-around specify whether to keep the value at the same sign as the original and how many most significant bits to set to one (according to the value of `nb`).

The library also offers limited-precision versions of fixed-point types, called `sc_fixed_fast<>` and `sc_ufixed_fast<>`, which have better performance than their counterparts `sc_fixed<>` and `sc_ufixed<>`. But for these types, the precision for intermediate results are limited to 53 bits on 32-bit platforms. These types offer the designer a trade-off in speed versus precision and may be used if the precision is sufficient.

4.4.7 Fixed-Point Parameters and Contexts

Sometimes it is necessary to vary the parameters of fixed-point objects dynamically. For instance, in the parameterizable methodology of chapter 6, we would like to make a single function usable for fixed-point types of different characteristics. For this purpose, SystemC introduces the notions of *fx-parameters* (class `sc_fxtype_params`) and *fx-context* (class `sc_fxtype_context`). Instead of the template types in section 4.4.6, which are *never* affected by fx-contexts, to create dynamically sized fixed-point objects we use the nontemplate types `sc_fix` and `sc_ufix`, and the limited-precision versions `sc_fix_fast` and `sc_ufix_fast`. Note the difference between `fixed` (template) and `fix` (nontemplate).

An fx-parameters object is used to specify the five fixed-point parameters (`wl`, `iwl`, etc.) either fully or partially (with restrictons on the allowable specified subset). Whenever a new fx-context C_1 corresponding to an fx-parameters object P_1 is brought into scope, the currently active fx-context C_0 (with parameters P_0), if present, is saved, and the parameter values of P_1 become the parameters for fixed-point objects created thereafter, until C_1 goes out of scope. When C_1 goes out of scope, the previously saved C_0 becomes active again. If P_1 has only partially specified parameters, those unspecified parameters will "inherit" their values from P_0. The exact syntax for manipulating fx-parameters and fx-context will become clear as we see examples in section 6.3.2.

4.5 Summary

In this chapter we have begun to see SystemC in action. On the one hand, we have learned how to use SystemC for classical RTL and behavioral-level modeling. On the other hand, through these examples we have gained a better feel for how various SystemC constructs work, such as signals, sensitivity, `wait()` and `wait_until()`. We have also examined the rich set of data-types that SystemC provides for hardware modeling. Equipped with this bag of tricks, we are now ready to climb the ladder of abstraction levels further up and explore functional and transaction-level modeling.

5

Functional Modeling

In the previous chapter we discussed how to describe the structure, interface, timing, and function of digital hardware in SystemC. In early stages of a design process however, especially if the partitioning of functionality into hardware and software is not yet determined, we are interested in only describing the functionality of a system's components. Details such as timing, fine-grain structure, or low-level communication protocols are best avoided at that point. This helps to keep the process of creating an *executable specification* as efficient as possible. Avoiding unnecessary implementation details will speed up both the modeling process itself as well as the simulation of the resulting specification. This approach is also advisable as it avoids overconstraining the set of implementation alternatives for a particular function.

We will start by taking a look at untimed functional models in the next section. These are often used to describe and analyze signal processing functionality such as filters, decoders, or demodulators. In section 5.2 we discuss how timing information (processing delays) can be introduced. This may be required if either modeling temporal relationships is needed in order to analyze the system behavior or if the functional model is used as a placeholder for its to-be-designed implementation. Finally, means to terminate a simulation—after all relevant data has been gathered—are described.

5.1 Untimed Functional Models – Dataflow

An effective model of computation (MOC) that is very commonly used at the untimed functional level is *dataflow modeling* (cf. section 3.4). Of course, it is possible to also use other models of computation at this level. We will not explore the use of other MOCs at this level within this chapter because their use at this level is less common. The most general deterministic, untimed model of computation is the *Kahn process network* [14] [15]. (Static [17] [18] and dynamic [5] dataflow models are special cases of Kahn process networks.) There, in the most general terms, behavior is described as functionality that maps sequences (streams) of input tokens onto sequences of output tokens [20]. This is modeled using processes communicating through FIFO channels that are accessed via blocking read and write operations. Algorithmic delays are represented by initial values stored in the FIFOs. The "*golden recipe*" to do this with SystemC is:

> *Use modules that contain `SC_THREAD` processes. Let these modules communicate through `sc_fifo` channels using the blocking `read()` and `write()` methods. Model initial values (algorithmic delays) either by writing to the FIFOs before calling `sc_start()` or by inserting modules that generate data before they consume. Stop the simulation by (a combination of) the following:*
>
> - *Simulate for a finite amount of time. (Requires at least one timed model to be present.)*
> - *Terminate processes (return from `SC_THREAD`) and stall the simulation through data backlog (eventually the simulation will stop due to the lack of events).*
> - *Call `sc_stop()` when a termination condition is fulfilled.*

Let us start with a simple example—an adder (example 5.1.1). The module (`DF_Adder`) is implemented as a class template such that it can be used for different data-types. It communicates with the outside world through FIFO ports. Its behavior is specified in a thread process (`SC_THREAD`) where the sum of the input tokens is written to the output port. Using thread processes is important as the FIFOs are accessed using blocking read and write methods. As a result, the synchronization is implicit. Write accesses block until space is available in the channel. Likewise, read accesses block until data is available. There is no need for explicit synchronization mechanisms; no token gets lost and processes are stopped and resumed automatically.

Example 5.1.1 *Dataflow adder*

```
// Simple dataflow adder. Works at least for builtin C types.
template <class T> SC_MODULE(DF_Adder) {
```

```
    sc_fifo_in<T> input1, input2;
    sc_fifo_out<T> output;
    void process() {
        while (1) output.write(input1.read() + input2.read());
    }
    SC_CTOR(DF_Adder) { SC_THREAD(process); }
};
```

Example 5.1.2 shows a constant generator (DF_Const) and a fork module (DF_Fork), which splits an incoming data stream into two identical value sequences. Such a fork is required when multiple modules need to consume the data generated by the same producer. Again, both modules are implemented as class templates, use thread processes, and communicate via blocking read and write operations.

Example 5.1.2 *Constant generator and fork module*

```
// Simple constant generator. Works at least for builtin C types.
template <class T> SC_MODULE(DF_Const) {
    sc_fifo_out<T> output;
    void process() { while (1) output.write(constant_); }
    SC_HAS_PROCESS(DF_Const); // needed as we do not use SC_CTOR

    // constructor w/ module name and constant
    DF_Const(sc_module_name N, const T& C) :
        sc_module(N), constant_(C) { SC_THREAD(process); }

    T constant_; // the constant value we write to the output
};

// This module forks a dataflow stream.
template <class T> SC_MODULE(DF_Fork) {
    sc_fifo_in<T> input;
    sc_fifo_out<T> output1, output2;
    void process() {
        while(1) {
            T value = input.read();
            output1.write(value);
            output2.write(value);
        }
    }
    SC_CTOR(DF_Fork) { SC_THREAD(process); }
};
```

As all modules are untimed, the simulation time will never advance other than in terms of delta-cycles. Hence, we cannot just tell the simulator to run for a certain amount of (simulated) time—it would never stop. Untimed simulations are typically run until a sufficient number of data samples have been produced. Example 5.1.3 shows a module (DF_Printer), which terminates its single thread process after a given number of iterations. This can cause the simulation to come to a halt as data traffic through this module stops and, as a result, other modules are stopped due to data backlog. Eventually, the SystemC simulator will stop due to the lack of events.

Example 5.1.3 *Printer module*

```
// Simple dataflow module that runs for a given number of
// iterations (c'tor argument) during which it prints the
// values read from its input on stdout.
template <class T> SC_MODULE(DF_Printer) {
    sc_fifo_in<T> input;
    SC_HAS_PROCESS(DF_Printer);

    // constructor w/ name and number of iterations
    DF_Printer(sc_module_name NAME, unsigned N_ITER) :
        sc_module(NAME), n_iterations_(N_ITER), done_(false)
        { SC_THREAD(process); }

    void process() {
        for (unsigned i=0; i<n_iterations_; i++) {
            T value = input.read();
            cout << name() << " " << value << endl;
      }
        done_ = true;
        return; // terminate process after given # iterations
    }

    // destructor: check whether we have actually read a
    // sufficient number of values when the simulation ends.
    ~DF_Printer() {
        if (!done_) cout << name() << " not done yet" << endl;
    }

    unsigned n_iterations_; // number of iterations
    bool done_; // flag indicating whether we are done
};
```

Figure 5.1 depicts an example system built up using the modules introduced so far. The constant generator feeds an adder that works as an integrator since its output is fed back into one of its inputs. In order to also print the output on the console we use a fork module to split the data stream.

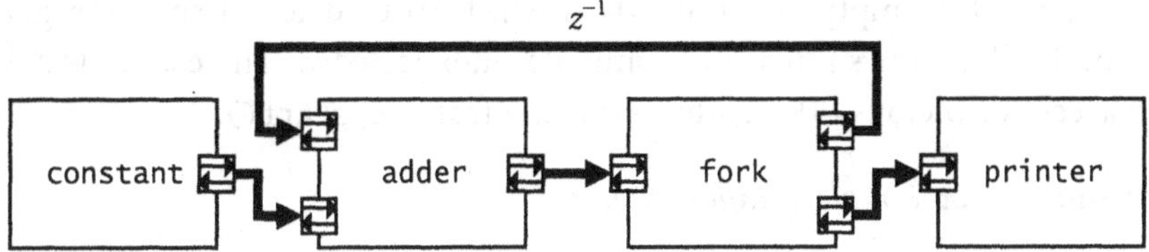

Figure 5.1: Simple dataflow system

Example 5.1.4 shows the corresponding source code. An initial value (indicated by a z^{-1} in figure 5.1) for the feedback channel is provided by explicitly writing to it before calling `sc_start()`. Note that the system would deadlock without this initial value, as the adder would never be able to successfully read from its first input.

We have defined a destructor for the printer module (`~DF_Printer()`, in example 5.1.3) as a means of checking whether the simulation has been stopped prematurely, i.e., before the desired number of samples has been produced. If we comment out the insertion of the initial value, then this will result in a message being printed at the end of the simulation (the system deadlocked before the printer could process the specified number of tokens).

Dataflow modeling in SystemC differs from most other approaches in that the user determines the FIFO sizes. Classical (pure) dataflow simulators tend to automatically size FIFO buffers during elaboration or even resize them during simulation runtime. In other words, queues are infinite and the system is nonlossy. It needs to be understood that the use of finite-sized FIFOs is mandatory in SystemC where the underlying simulation semantics are event-driven. It can easily be seen that connecting the `DF_Const` module (example 5.1.2) to an automatically resizing FIFO would result in unbounded memory consumption.

Missing (initial) data values are not the only cause that can stall a SystemC dataflow simulation. The production and consumption of data need to be generally balanced. That is, a producer and a consumer connected to a single FIFO need to move the same number of tokens—at least on average over time (the bigger the variations, the more buffer storage is required). Assume that the `DF_Fork` instance in figure 5.1 produces two values on each output for every data sample consumed. Sooner or later the simulation would stall with the fork module being stuck attempting to write to its first output port, as the rates of data production and consumption do not match for the feedback loop.

A quest for empty and full FIFOs will help to detect a smoking gun should a dataflow simulation terminate unexpectedly. The easiest way is to access suspicious FIFOs after returning from sc_start().

Example 5.1.4 *Simple dataflow example*

```
int sc_main(int, char**)
{
    // module instances
    DF_Const<int> constant("constant", 1);
    DF_Adder<int> adder("adder");
    DF_Fork<int> fork("fork");
    DF_Printer<int> printer("printer", 10);
    // fifos
    sc_fifo<int> const_out("const_out", 5);
    sc_fifo<int> adder_out("adder_out", 1);
    sc_fifo<int> feedback("feedback", 1);
    sc_fifo<int> to_printer("to_printer", 1);

    // initial values; forget about this and the
    // system will deadlock
    feedback.write(42);

    // interconnect
    constant.output(const_out);
    adder.input1(feedback);
    adder.input2(const_out);
    adder.output(adder_out);
    fork.input(adder_out);
    fork.output1(feedback);
    fork.output2(to_printer);
    printer.input(to_printer);

    // Start simulation w/o time limit. The simulation will stop
    // when there are no more events. Once the printer module
    // has terminated the complete simulation will eventually
    // come to a halt (after all fifos have been filled up)
    sc_start(-1);

    return 0;
}
```

5.2 Timed Functional Models

Introducing a notion of time into our untimed functional models is simple; we annotate processing delays using wait(sc_time) (cf. section 2.7.2).

We could, for instance, change the constant generator (example 5.1.2) such that the generation of each new value takes 200 nanoseconds.

```
void process() {
    while (1) {
        wait(200, SC_NS);
        output.write(constant_);
    }
}
```

Similarly, we could add some processing delay to the adder (example 5.1.1):

```
void process() {
    while (1) {
        T data = input1.read() + input2.read();
        wait(200, SC_NS);
        output.write(data);
    }
}
```

This example also gives us an indication of why the use of thread processes (SC_THREAD) is advisable. Not only is it next to impossible to maintain blocking read and write semantics for method processes (SC_METHOD) but it also becomes quite challenging to add delay information to nontrivial method processes.

In the world of SystemC, timed and untimed models can peacefully coexist and interact. The underlying event-driven simulation semantics plus the fact that we use finite size FIFOs make it possible to connect them in simple (yet deterministic) ways. One can even mix FIFO- and signal-based communication as shown in example 5.2.1. There, a controller that is implemented as a finite state machine (FSM) in a hardware-like coding style could be used to generate the coefficient value. (Note that the use of such a coefficient module controlled by an sc_signal channel means that the system no longer uses a pure Kahn process network model of computation.)

Example 5.2.1 *Mixing FIFO and signal I/O*

```
// This module multiplies a value sequence with a coefficient
// that is read from a (signal) input port.
template <class T> SC_MODULE(DF_CoeffMul) {
    sc_fifo_in<T> input;
    sc_fifo_out<T> output;
    sc_in<T> coefficient;
    void process() {
        while (1)
```

```
            output.write(input.read() * coefficient.read());
    }
    SC_CTOR(DF_CoeffMul) { SC_THREAD(process); }
};
```

It should be noted that FIFO sizes play a key role. Not only do they determine whether a system deadlocks (see last section)—they can also have an impact on the system behavior (value sequences) and the simulation speed. (This effect is attenuated as the interaction between timed and untimed parts increases.) Assume a scenario where the input of the coefficient multiplier (example 5.2.1) is connected to the untimed constant generator (example 5.1.2) while its output fuels some timed functional model. The sizes of input and output FIFOs determine how many samples are produced at `t=0` and how often processes have to be activated for that. Larger FIFO sizes will result in a smaller number of activations (context switches) and, hence, in better simulation speed. The size of the output FIFO influences how many values can be written in the beginning, that is, how many input values are multiplied with the value the coefficient has at `t=0`. Thus, a different size may result in a different value sequence being generated.

5.3 Stopping a Dataflow Simulation

In this section we will briefly review three ways to stop a dataflow simulation. We have learned in the first section of this chapter that using `sc_start()` with a positive, nonzero argument will not lead to the desired result in a purely untimed scenario. We used process termination and hoped for data backlog to eventually result in no processes being able to run, thus causing the simulation to stop.

```
void process() {
    for (unsigned i=0; i<n_iterations_; i++) ...
    ...
    return; // terminate process after given # iterations
}
```

This approach has some shortcomings, though. First, it is not guaranteed to be successful. The process may be in some dead branch of the system that is no longer supplied with data so that termination will not result in the necessary back pressure. Furthermore, we could have multiple instances of `DF_Printer` in our system and might be interested in letting the simulation run until all of them have reached their respective number of iterations. In fact, this type of logical AND-combination of exit conditions is typical for many simulation scenarios.

Example 5.3.1 shows a modified version of the DF_Printer module that does not terminate its thread process. Instead, it just indicates the fact that it is done via a Boolean signal (this is another case of mixed dataflow/signal communication) and continues to consume input values in order to avoid stalling the simulation due to data backlog.

The Terminator module (cf. example 5.3.1) can be connected to an arbitrary number of such Boolean signals. (Note here the use of the multiport sc_port<sc_signal_in_if<bool>,0> rather than sc_in<bool>. We will see more examples of multiports in section 8.8.) It terminates the simulation by calling sc_stop() as soon as all input signals assume the value true. Multiple Terminator instances can be used in a single simulation, effectively resulting in specifying the exit condition of a simulation as a logical sum of products (canonical disjunctive form).

Example 5.3.1 *Using modules to terminate a simulation*

```
// Simple dataflow module that runs for a given number of
// iterations  (template argument) during which it prints
// the values read from its input on stdout.
template <class T, unsigned n_iterations> SC_MODULE(DF_Printer) {
    sc_fifo_in<T> input;
    sc_out<bool> done;
    SC_CTOR(DF_Printer) {
        SC_THREAD(process);
        done.initialize(false);
    }
    void process() {
        for (unsigned i=0; i<n_iterations; i++)
            cout << name() << " " << input.read() << endl;
        done = true;
        while(1) input.read(); // avoid data backlog
    }
};

// This module terminates the simulation when all signals
// connected to its input are set to 'true'.
SC_MODULE(Terminator) {
    sc_port<sc_signal_in_if<bool>, 0> inputs;
    SC_CTOR(Terminator) {
        SC_METHOD(arnold);
        sensitive << inputs;
    }
    void arnold() { // terminate when all inputs are 'true'
        for (unsigned i=0; i<inputs.size(); i++)
```

```
            if (inputs[i]->read() == false) return;
        // we only get to this point if all inputs are 'true'
        sc_stop();
    }
};
```

If the simulation contains timed modules then using `sc_start()` with a positive argument works—except for those cases where time is not advanced because the simulation turns out to degenerate into an endless sequence of delta-cycles. (Possible causes are combinational feedback loops or the presence of untimed dataflow subgraphs that do not depend on the activation of a timed module.) Still, approaches similar to the `Terminator` module may be advisable as it is often hard to predict the minimal (simulated) time it takes to produce certain data. Combinations of a given maximum simulation time and data-dependent exit conditions can often be the best solution.

5.4 Summary

In this chapter we have seen how `SC_THREAD` processes and `sc_fifo` channels can be used to write untimed functional models in a dataflow-oriented way. The main difference to "classical" dataflow approaches is the use of finite sized FIFO channels. Functional modeling in SystemC goes beyond this, though. Processing delays are introduced easily via `wait(sc_time)`. Furthermore, a variety of communication schemes—without being restricted to point-to-point protocols—can be used. As an example, we have seen FIFO and signal-based communication peacefully coexist when we discussed ways to terminate a simulation after all relevant data has been generated.

6

Parameterized Modules and Channels

6.1 Introduction

One of the most powerful features within SystemC is its ability to support parameterized design. Because designers often invest considerable effort in the development of modules and channels, it pays to design them so that they can be reused as much as possible. SystemC provides several different ways to create parameterized modules and channels, each providing unique modeling capabilities. For example, in SystemC it is possible to create a FIFO module or channel that stores elements whose data-type is parameterized. This can't be done in traditional design languages such as VHDL and Verilog.

This chapter first outlines the different techniques for constructing parameterized modules and channels in SystemC, and then provides some design examples and guidelines for using parameters. Lastly we explore techniques for protecting intellectual property within parameterized libraries.

6.2 Forms of Parameterization

Before discussing the techniques for using parameters within SystemC, it is useful to outline some of the different aspects of a design that might be parameterized in SystemC. A nonexhaustive list of *what* can be parameterized in SystemC includes:

- *Parameterized values within a design.* For example, the coefficients for a finite impulse response (FIR) filter or the constant data values within a read-only memory might be specified using parameterized values.

- *Parameterized types and type attributes.* For example, an FIR filter might have a template parameter specifying whether a floating-point or fixed-point type is used, or it might have parameters specifying fixed-point type attributes such as total width and the number of integer bits.

- *Parameterized design structure.* For example, the structure of a parameterized width carry-lookahead adder might be specified using if and for statements that conditionally instantiate parts of the adder based on the value of the width parameter.

When are parameter values resolved? You have exactly three choices in SystemC, as shown below. Depending on when parameters need to be resolved, different techniques are available in SystemC for specifying parameters.

- *At compile time.* The parameter values or types are resolved when the C++ compiler is run. To resolve parameters at compile time, use C++ template parameters.

- *At elaboration time.* The parameter values are resolved when the SystemC design hierarchy is constructed prior to the start of simulation. (i.e., after `sc_main()` is called but before `sc_start()` is called.) To resolve parameters at elaboration time, use C++ constructor arguments for modules and channels.

- *At simulation time.* The parameter values are resolved as the design is simulated, and may change as simulation occurs. To resolve parameters at simulation time in SystemC, use a channel such as `sc_signal<T>` to transport the updated parameter values during simulation to the places where it is needed. Alternatively, you can create a data member within the module or channel to represent the parameter, and then provide a public method that allows the data member's value to be set dynamically as simulation proceeds.

As we shall see in the rest of this chapter, understanding the correspondence between the time of parameter resolution and the different techniques for specifying parameters is a key part of creating parameterized modules and channels. We will now present a variety of modeling examples that demonstrate the different parameterization techniques.

6.3 Parameterized Designs Examples

6.3.1 Compile-Time Resolution

As mentioned in the previous section, compile time resolution of parameter values is achieved with the use of C++ template parameters. With template parameters, users can specify parameter arguments to the module or channel that allow both type arguments and integer arguments to be set at the point of module instantiation. To demonstrate compile time resolution of parameters, the following finite impulse response (FIR) filter design uses template parameters to specify the number of taps and the numeric data-type used within the FIR.

```
#define SC_INCLUDE_FX
#include <systemc.h>

// class template "fir"
//
// Template Parameters:
//   class T - specifies the data-type used within the FIR
//     T must be a numeric type that supports:
//      operator==(const T&)
//      operator=(int)
//      operator+=(const T&)
//      operator*(const T&)
//   unsigned N - specifies the number of taps in FIR
//     N must be greater than zero
//
// Constructor parameters:
//   sc_module_name name - specifies instance name
//   const T* coeffs - pointer to coefficient array
//     coeffs array must contain N coefficients

template <class T, unsigned N> class fir: public sc_module {
public:
    sc_in<bool> clock;
    sc_in<T> in;
    sc_out<T> out;

    SC_HAS_PROCESS(fir);

    fir(sc_module_name name, const T* coeffs) :
        sc_module(name), _coeffs(coeffs)
```

```
    {
        assert(N > 0);
        SC_METHOD(main);
        sensitive << clock.pos();

        for (int i=0; i < N; i++)
            _delay_line[i] = 0;
    }

private:
    T _delay_line[N];
    const T* _coeffs;

    void main() {
        // shift samples within delay line
        for (int j=N-1; j > 0; j--)
            _delay_line[j] = _delay_line[j-1];

        // read new data sample
        _delay_line[0] = in.read();

        // compute fir output
        T sum = 0;
        for (int i=0; i < N; i++)
            sum += _delay_line[i] * _coeffs[i];

        out.write(sum);
    }
};
```

The following code shows how this FIR module can be instantiated within a design:

```
// simple stimulus generator

template <class T> class stimulus : public sc_module {
public:
    sc_out<T> out;

    SC_HAS_PROCESS(stimulus);

    stimulus(sc_module_name name) : sc_module(name) {
        SC_THREAD(main);
    }

    void main() {
        out.write(0);
        wait(10, SC_NS);
        out.write(2);
```

```
        wait(1, SC_NS);
        out.write(0);
    }
};

// simple response logger

template <class T> class response : public sc_module {
public:
    sc_in<T> in;

    SC_HAS_PROCESS(response);

    response(sc_module_name name) : sc_module(name) {
        SC_METHOD(main);
        sensitive << in;
    }

    void main() {
        cout << "at time: " << sc_time_stamp() << " output: "
             << in.read() << endl;
    }
};

int sc_main (int argc , char *argv[])
{
    // to use a fixed-point type, uncomment next line
    // and comment out the line after it:
    // typedef sc_fixed<8,5> fir_T;
    typedef double fir_T;

    const fir_T coeffs[] = {1.1111, 2.2222, 3.3333, 4.4444};
    const unsigned taps = sizeof(coeffs) / sizeof(coeffs[0]);

    sc_clock clock("c1", 1, SC_NS);
    sc_signal<fir_T> fir_in;
    sc_signal<fir_T> fir_out;

    fir<fir_T, taps> fir1("fir1", &coeffs[0]);
    fir1.clock(clock);
    fir1.in(fir_in);
    fir1.out(fir_out);

    stimulus<fir_T> stim1("stim1");
    stim1.out(fir_in);

    response<fir_T> resp1("resp1");
    resp1.in(fir_out);
```

```
    sc_start(100, SC_NS);
    return 0;
}
```

Note that the FIR data-type and number of taps are specified at the point where the FIR is instantiated, not within the FIR module itself. This allows the data-type and number of taps to be customized each time the FIR is instantiated. To demonstrate the power of this technique, the data-type can be changed from a floating point representation to an 8-bit fixed-point representation by uncommenting the line

```
typedef sc_fixed<8,5> fir_T;
```

and commenting out the line

```
typedef double fir_T;
```

Such techniques make it easy to explore the impact that the use of fixed-point arithmetic will have on a design. (For more information on the SystemC fixed-point data-types, see section 4.4.6.)

Because the FIR data-type and number of taps parameters are resolved at compile time, it is still possible for compilers or synthesis tools to perform many optimizations on this FIR design. For example, if we use the `sc_fixed<8,5>` data-type, a C++ compiler or a hardware synthesis tool should be able to recognize that only eight bits of storage are needed for each data sample, and since this FIR instantiation only uses four taps, only four bytes of storage are needed for the delay line.

6.3.2 Elaboration-Time Resolution

Parameter resolution at elaboration time in SystemC is achieved via the use of constructor arguments to modules and channels. In general, it is straightforward to write models that use such constructor arguments. The following FIR design example is similar to the previous design example, except that it uses constructor arguments to the FIR module to specify the attributes used for the fixed-point types within the FIR. Because the attributes are only resolved at elaboration time, this design example allows the user to specify the fixed-point attributes as command line arguments when this design is simulated. For example, the command line arguments `"8 5"` result in the same fixed-point attributes as in the above example (8 total bits used for the fixed-point numbers, with 5 integer bits), while the arguments `"32 16"` will result in fixed-point precision that is almost as precise as the float data-type.

```
#define SC_INCLUDE_FX
#include <systemc.h>

// module "fir"
// Constructor parameters:
//   sc_module_name name - specifies instance name
//   const double* coeffs - pointer to coefficient array
//     coeffs array must contain "n" coefficients
//   unsigned w - total bit width of fixed pt type used
//   unsigned I - number of integer bits of fixed pt type
//   unsigned n - number of taps in FIR

class fir : public sc_module {
public:
    sc_in<bool> clock;

    // For the sake of simplicity, the "sc_fix" data-type
    // is only used inside this module, and all external
    // interfaces rely on "double"

    sc_in<double> in;
    sc_out<double> out;

    SC_HAS_PROCESS(fir);

    fir(sc_module_name name, const double* coeffs, unsigned w,
        unsigned i, unsigned n) :
        sc_module(name), _w(w), _i(i), _n(n)
    {
        assert(n > 0); assert(w > 0);
        SC_METHOD(main);
        sensitive << clock.pos();

        // see discussion below for explanation of next line
        sc_fxtype_context c1(sc_fxtype_params(_w, _i));
        _delay_line = new sc_fix[_n];
        _coeffs = new sc_fix[_n];

        // copy input coeffs array and convert to sc_fix type
        for (unsigned j=0; j < _n; j++)
            _coeffs[j] = coeffs[j];
    }

    ~fir() { delete[] _delay_line; delete[] _coeffs; }

private:
    sc_fix *_delay_line;
    sc_fix *_coeffs;
    const unsigned _w, _i, _n;
```

```
    void main() {
        // shift samples in delay line
        for (int j=_n-1; j > 0; j--)
        _delay_line[j] = _delay_line[j-1];

        // read new data sample
        _delay_line[0] = in.read();

        // compute fir output
        sc_fix sum(_w, _i);
        sum = 0;
        for (unsigned i=0; i < _n; i++)
            sum += _delay_line[i] * _coeffs[i];

        out.write(sum);
    }
};


// template class "stimulus" is reused from first example
// template class "response" is reused from first example

int sc_main (int argc , char *argv[])
{
    const double coeffs[] = {1.1111, 2.2222, 3.3333, 4.4444};
    const int taps = sizeof(coeffs) / sizeof(coeffs[0]);
    int w = 8; // total width of fixed pt numbers
    int i = 5; // integer bits of fixed pt numbers

    if (argc > 1) w = atoi(argv[1]);

    if (argc > 2) i = atoi(argv[2]);

    sc_clock clock("c1", 1, SC_NS);
    sc_signal<double> fir_in;
    sc_signal<double> fir_out;

    fir fir1("fir1", &coeffs[0], w, i, taps);
    fir1.clock(clock);
    fir1.in(fir_in);
    fir1.out(fir_out);

    stimulus<double> stim1("stim1");
    stim1.out(fir_in);

    response<double> resp1("resp1");
    resp1.in(fir_out);
```

```
    sc_start(100, SC_NS);
    return 0;
}
```

One aspect in the example above that deserves explanation is the line within the FIR module that refers to the sc_fxtype_context class (see section 4.4.7). Because the _delay_line and _coeffs arrays within the FIR module are constructed as arrays of sc_fix elements, the default constructor for sc_fix (i.e., the constructor that takes no arguments) is called for each element in the array. However, we wish to use the fixed-point attributes specified by w and i when constructing the elements within these arrays. To do so, this example creates a sc_fxtype_context object to cause these fixed-point attributes to be used during construction.

As this example illustrates, the ability to set parameters at elaboration time can be used as a powerful way to perform design exploration without the need for recompilation. The same techniques used within this design example could be applied to the design of a larger communications system where the objective was to determine the optimal fixed-point attributes for several parts of the design. A series of simulations could then be run in which the fixed-point attributes are modified to assess the overall effect on the system, all without the need to perform any recompilation of the design.

Keep in mind, however, that constructor arguments cannot be used to specify C++ types in the way that template parameters are used. With the current example above, we see that the FIR module must always use the sc_fix data-type, but that the data-type can be configured via the module constructor arguments.

The next design example illustrating the use of elaboration time parameter resolution demonstrates how parameters can be used to influence the structure of the design at elaboration time. The first module reg within this example models a simple register with an integer input and output that is sensitive to the rising edge of the clock. Next, we implement a shift register module named shiftreg that is built from the reg module. The shift register module has a constructor argument that specifies the length of the shift register, and within the constructor for shiftreg we iterate through a loop and dynamically instantiate the appropriate number of reg instances. In addition, we dynamically instantiate signals to connect the reg instances and also connect the first and last reg instances within the shift register to the input and output ports of the shift register. Although this example may look different from previous structural SystemC examples, it is really just using the normal mechanisms for creating design structure in SystemC. The main difference is that we are now manipulating pointers to modules and signals rather than working with these objects directly.

```
#include <systemc.h>

// reg: a simple register that stores an "int"

class reg : public sc_module {
public:
    sc_in<bool> clock;
    sc_in<int> in;
    sc_out<int> out;

    SC_HAS_PROCESS(reg);

    reg(sc_module_name name) : sc_module(name) {
        SC_METHOD(main);
        sensitive << clock.pos();
    }

    void main() { out = in.read(); }
};

// shiftreg: a shift register that stores "ints".
//   Overall shift register length is set via
//   its constructor argument

class shiftreg : public sc_module {
public:
    sc_in<bool> clock;
    sc_in<int> in;
    sc_out<int> out;

    shiftreg(sc_module_name name, unsigned len)
      : sc_module(name)
    {
        char buf[10];

        // "prev" points to the signal connected to the
        // output of the previous "reg" instance

        sc_signal<int>* prev = 0;

        // loop to create and connect all the "reg" instances:

        for (unsigned i=0; i < len; i++)
        {
            sprintf(buf, "r%d", i);
```

```
            // instantiate the "reg" instance
            reg* regp = new reg(buf);

            // connect the "clock" port of every instance
            regp->clock(clock);

            // if this is the first instance, connect the "reg"
            // input port to the "shiftreg" input port, else
            // connect the "reg" input port to the output
            // signal of the previous "reg" instance
            if (i == 0)
                regp->in(in);
            else
                regp->in(*prev);

            if (i < len - 1) {
                // for each "reg" instance except the last,
                // create a new signal and attach it to the
                // output port on "reg"
                sc_signal<int>* sigp = new sc_signal<int>;
                regp->out(*sigp);
                prev = sigp;
            }
            else {
              // connect the output port of the last "reg"
              // instance to the output port of "shiftreg"
               regp->out(out);
            }
        }
    }

    ~shiftreg()
    {
        // a robust version would delete all the dynamically
        // allocated signals and modules
    }
};


// template class "stimulus" is reused from first example
// template class "response" is reused from first example

int sc_main (int argc , char *argv[])
{
    int w = 10;

    if (argc > 1) w = atoi(argv[1]);
    if (w < 1) w = 1;
```

```
    if (w > 50) w = 50;

    sc_clock clock("c1", 1, SC_NS);
    sc_signal<int> shiftreg_in;
    sc_signal<int> shiftreg_out;

    shiftreg sr1("sr1", w);
    sr1.clock(clock);
    sr1.in(shiftreg_in);
    sr1.out(shiftreg_out);

    stimulus<int> stim1("stim1");
    stim1.out(shiftreg_in);

    response<int> resp1("resp1");
    resp1.in(shiftreg_out);

    sc_start(100, SC_NS);
    return 0;
}
```

The example above is quite simple, but it gives some idea of the power of using parameters to control the algorithmic generation of structure within SystemC. Such techniques become particularly useful when creating large or complex design structures that can be more easily created via an algorithm rather than explicitly constructed.

6.3.3 Simulation-Time Resolution

In addition to resolving parameters at compile and elaboration time, it is also possible to set parameter values as the simulation proceeds within SystemC. There are two simple techniques that can be used to do this.

The first technique is to simply use an `sc_signal<>` channel instance and `sc_signal` ports (i.e., `sc_in<>`, `sc_out<>`) to transport the parameter to the places in the design where it is used. Then, as simulation proceeds and the parameter is assigned new values, all of the locations where the value is consumed will be automatically updated. This technique is illustrated in example 5.2.1 within chapter 5.

The second technique is to make the parameter a data member of the module or channel, and then to provide a module or channel method that allows the parameter to be set or reset. To illustrate this technique, we can make a small modification to the first FIR example within this chapter by making the coefficient array a data member within the FIR module, and then by providing the `set_coeff()` method to allow elements within the array to be set. This new design is shown below. With this new design it is possible to set the FIR's coefficients at any time during execution of the design.

```
template <class T, int N> class fir : public sc_module {
public:
    sc_in<bool> clock;
    sc_in<T> in;
    sc_out<T> out;

    SC_HAS_PROCESS(fir);

    fir(sc_module_name name) : sc_module(name) {
        SC_METHOD(main);
        sensitive << clock.pos();

        for (int i=0; i < N; i++)
            _delay_line[i] = _coeffs[i] = 0;
    }

    void set_coeff(unsigned i, T val) {
        if (i < N)
            _coeffs[i] = val;
    }

private:
    T _delay_line[N];
    T _coeffs[N];

    void main(); // same as in original FIR example
};
```

6.4 Guidelines for Using Parameters in SystemC

In the previous sections we have seen that SystemC supports several forms of parameterized models: template parameters for compile time resolution, constructor arguments for elaboration time resolution, and the use of `sc_signal<>` or user-defined module/channel methods for simulation time resolution.

Which form should you use in a particular case? If you know (for design reasons) that a parameter needs to be able to be set at elaboration time or simulation time, then you should use the appropriate parameterization technique as outlined above. Otherwise, it is usually best to create SystemC models such that parameters are resolved at the earliest possible time. In other words, it is usually best to prefer compile time resolution over elaboration time resolution, and elaboration time resolution over simulation time resolution. The reasons for this are:

- Design errors involving parameters can be caught earlier.
- Design errors involving parameters may be more thoroughly detected. For example, type incompatibilities involving parameter values can often be detected by the C++ compiler when compile time resolution is used.
- C++ compilers and synthesis tools may be able to perform better code optimization if parameters are resolved at compile time.
- Some synthesis and analysis tools may not fully support elaboration time and simulation time resolution of parameter values.

However, there are some important exceptions and clarifications to the above rules:

- As discussed in the next section, it can sometimes be difficult to protect intellectual property within code that uses C++ templates. In these cases it may be desirable to avoid the use of template parameters and instead rely on constructor arguments so that source code representing intellectual property does not need to be delivered to customers.
- SystemC 2.0 does not yet allow simulation time modification of design structure. In other words, modules and channels and their interconnections cannot be modified during simulation. Instead, design structure can be parameterized and created at elaboration time using constructor arguments as shown earlier in this chapter.
- Don't abuse constructor arguments by using them to build communication links between modules that will then be used during simulation. For example, don't pass pointers or callback functions via constructor arguments and then use these to pass information between modules once simulation starts. Instead, use normal SystemC ports to perform such communication so that the communication between modules is apparent to readers of your code.

There are a number of caveats in using C++ templates. Despite the benefits mentioned above concerning earlier error detection and potentially more efficient code generation, you should be aware of the following:

- Code that relies on template parameters usually needs to be included in the "`.h`" file rather than the "`.cpp`" file. This can make intellectual property protection difficult, and depending on your C++ compiler can sometimes result in long compile times, unnecessarily large object files and executables, and strange and confusing error messages. (After a while, you may develop a fine appreciation of the differences between various C++ compilers.) One common approach to reducing the size of "`.h`" files and executable files in these cases is to "factor out" common code into nontemplated base classes or functions. This "factored out" code can then be placed into a "`.cpp`" file, thus reducing the size of the "`.h`" file and avoiding many of the problems associated with templates.

- The C++ language standard requires that template parameters can only be types or simple integer arguments. You cannot pass objects such as strings, fixed-point numbers, or arrays as template parameters; you must use constructor arguments for these instead. As a case in point, in the first FIR example in this chapter we had to pass the constant coefficient array as a constructor argument rather than a template parameter. (Some compilers may support passing pointers as template parameters, and the C++ standard indeed allows this. But we do not recommend it since you get little benefit over constructor arguments.)

- Don't lose sight of the fact that parameterized models that contain parameterized types within them can only be created with template parameters. For instance, to create a register module with ports that are declared as `sc_in<sc_uint<N> >` and `sc_out<sc_uint<N> >` you must make the register module itself templated with `N` as a template parameter. You cannot specify `N` via a module constructor argument. In contrast, some types such as `sc_fix` are not parameterized, but do allow the type to be configured at the time that `sc_fix` objects are constructed. You can use constructor arguments (in addition to template parameters) to perform such type configuration.

Also remember these general C++ rules for creating parameterized code:

- Document the interface for your parameterized modules and channels, including the proper use of parameters. Specify the required methods and operators for template type parameters, and specify constraints on parameter values and combinations.

- Include code in the module to check parameter values to make sure they are valid.
- Include code in the module to check that the combination of parameters is valid (each individual parameter may be valid but the combination may have to be rejected).
- Include default values for template parameters and constructor arguments, but only when the default value would be used in most cases.

When creating parameterized modules and channels, it is usually wise to avoid using the `SC_MODULE` and `SC_CTOR` macros. Instead, use normal C++ syntax without these macros, as illustrated in the code examples within this chapter. Note that modules or channels that do not use `SC_CTOR` must use the `SC_HAS_PROCESS` macro if the module or channel contains any processes. Note also that the first argument to any constructors should be an `sc_module_name` object, passed by value (note the absence of `&`), and that this argument is passed on to the `sc_module` or `sc_channel` base class constructor.

SystemC allows you to add attributes to objects such as channels and modules. (This is done via the functions `sc_object::add_attribute()` and `sc_object::get_attribute()`, which are not yet part of the official SystemC standard.) It is important to clearly distinguish between parameters and attributes. Parameters can be read by a module or channel, and can alter the module's behavior. Attributes should not be read by the module itself, and should not alter the module's behavior. Attributes are commonly used as a way to annotate information (e.g., placement data) for use by downstream implementation tools.

All of the guidelines in this section are only intended to provide a general set of rules. If you are using specific synthesis or implementation tools for your SystemC designs, also consult the tool documentation for more detailed coding guidelines.

*6.5 Protecting Intellectual Property within Parameterized Libraries

In some cases parameterized modules or channels represent a company's intellectual property that is delivered to designers outside of the company.

An important question then arises: How can companies deliver parameterizable modules or channels in secure form in SystemC? Companies typically wish to deliver such designs in compiled form, since this prevents users from easily observing the internal structure of designs. If such designs do not have unresolved template parameters within them, then it is straightforward to compile the needed files and deliver the object code to customers. There are even ways to strip out unneeded symbols from object files so that only those symbols required for linking with the customer's code are present. However, if the design has template parameters that must be set by the end user, then C++ source code for parts of the design that are dependent on the template parameters usually must also be delivered to the customer, since the C++ compiler requires source code when it instantiates templates.

If the templated module or channel that you deliver to the customer will only be instantiated with a small number of different parameter combinations, then the simplest strategy is to use C++ *explicit template instantiation* [33] to create object code for those specific parameter profiles. This allows you to not provide source code for the module or channel implementation to the customer. The following is a simple example that demonstrates explicit instantiation of a templated module.

```
// This is the "cpu.h" header file provided to customer

#include <systemc.h>

// A simple CPU with parameterized bus width & cache size

template <unsigned W> class cpu : public sc_module
{
public:
    sc_in<bool> clock;
    sc_out<sc_uint<W> > addr;
    sc_inout<sc_uint<W> > data;

    SC_HAS_PROCESS(cpu);
    cpu(sc_module_name nm, unsigned cache_size);

private:
    void main();
    unsigned _cache_size;
};
// end of "cpu.h"
```

```
// This is the "cpu.cpp" file. Only the "cpu.o" file
// is provided to the customer

template <unsigned W>
cpu<W>::cpu(sc_module_name nm, unsigned cache_size) :
    sc_module(nm),
    _cache_size(cache_size)
{
    SC_METHOD(main);
    sensitive << clock.pos();
}

template <unsigned W>
void cpu<W>::main()
{
    // here's where the "secret" functionality goes
    static int i;
    addr = i++;
    data = _cache_size + i;
}

// explicitly instantiate cpu for 16, 32, 64 bit bus widths:

template class cpu<16>;
template class cpu<32>;
template class cpu<64>;

// end of "cpu.cpp"
```

In this example the customer will be able to instantiate the cpu module with bus widths of 16, 32, and 64. If he tries to instantiate other bus widths, he will get an error at link time.

If the above approach won't work in a particular case because a module or channel may be instantiated with too many unique parameter combinations, another effective way to address this issue is to "factor out" any functionality that needs to be kept confidential and that is not directly dependent on the template parameters into a nontemplated base class for the module, or into nontemplated functions that are used by the module. This code can then be compiled and shipped in object form rather than in source form.

The final way to address this issue is illustrated in the example below. The basic strategy that is used here is to create a templated wrapper module (called wrapper within this example) that presents the "expected" interface to the user. In this example the wrapper module has a template parameter W that specifies the width used within its implementation, and it presents input and output ports whose underlying type are `sc_uint<W>`. Within this module we then have an embedded module `secret` that contains the functionality that we wish to protect by delivering in object form. One might be tempted to declare the input and output ports of `secret` as having type `sc_uint<W>`, but we can't do that since it would then make `secret` dependent on a template parameter and then we would need to deliver source code for it. One might next be tempted to make the type of `secret`'s ports be `sc_uint_base` (i.e., the nontemplated base class for `sc_uint<>`), but this is not possible because these types are not supposed to be used directly within SystemC designs, and also because there is no way to specify constructor arguments (such as the width of the data-type) to types such as `sc_uint_base` when they are used as the underlying types for signals and ports.

To address these issues, this example leverages data-types that are intended for use within SystemC designs and that provide a way to specify constructor arguments when used as the underlying types for signals and ports. These data-types are `sc_ufix` and `sc_fix`. Thus, in this example we implement the `secret` module using ports of type `sc_ufix` and then convert these types to `sc_uint<>` within the wrapper module.

```
// This is the header file delivered to the customer
#define SC_INCLUDE_FX
#include "systemc.h"

class secret : public sc_module
{
private:
    sc_fxtype_context _context; // must declare before ports
public:
    sc_in<sc_ufix> in1;
    sc_in<sc_ufix> in2;
    sc_out<sc_ufix> out1;

    SC_HAS_PROCESS(secret);

    secret(sc_module_name nm, unsigned w);

private:
    void main();
    unsigned _w;
    void *secret_data; // main() can save secret data here
};
```

```
template <unsigned W>
class wrapper : public sc_module
{
public:
    SC_HAS_PROCESS(wrapper);
    sc_in<sc_uint<W> > in1;
    sc_in<sc_uint<W> > in2;
    sc_out<sc_uint<W> > out1;

    wrapper(sc_module_name nm)
        : sc_module(nm),
          mult1("mult1", W),
          _context(sc_fxtype_params(W, W))
    {
        SC_METHOD(in1_method);
        sensitive << in1;
        SC_METHOD(in2_method);
        sensitive << in2;
        SC_METHOD(out1_method);
        sensitive << out1_sig;
        _context.end();

        mult1.in1(in1_sig);
        mult1.in2(in2_sig);
        mult1.out1(out1_sig);
    }

private:
    secret mult1;
    sc_fxtype_context _context; // must declare before signals
    sc_signal<sc_ufix> in1_sig, in2_sig, out1_sig;

    void in1_method()
    { sc_ufix t(W,W); t = in1.read(); in1_sig = t;}
    void in2_method()
    { sc_ufix t(W,W); t = in2.read(); in2_sig = t;}
    void out1_method()
    { sc_uint<W> t; t = out1_sig.read(); out1 = t; }
};


// This is the "secret" functionality within a ".cpp" file
// that is compiled and delivered to the customer as a ".o"

secret::secret(sc_module_name nm, unsigned w)
    : sc_module(nm), _context(sc_fxtype_params(w, w)), _w(w)
{
    SC_METHOD(main);
```

```
        sensitive << in1 << in2;
        _context.end();
    }

    void secret::main() { out1 = in1.read() * in2.read() + _w; }

    // This is how the customer might instantiate the
    // module (here named "wrapper") provided by the IP vendor

    // template class "stimulus" is reused from first example
    // template class "response" is reused from first example

    int sc_main (int argc , char *argv[])
    {
        const unsigned width = 16;
        typedef sc_uint<width> type;

        sc_signal<type> mult_in;
        sc_signal<type> mult_out;

        wrapper<width> mult1("mult1");
        mult1.in1(mult_in);
        mult1.in2(mult_in);
        mult1.out1(mult_out);

        stimulus<type> stim1("stim1");
        stim1.out(mult_in);

        response<type> resp1("resp1");
        resp1.in(mult_out);

        sc_start(100, SC_NS);
        return 0;
    }
```

Note that this last approach for protecting intellectual property within templated modules should only be used when the first two approaches outlined above cannot be used. This is because the last approach is both significantly more complex than the first two and also because it may be significantly slower due to the additional processes and the use of the `sc_fix` and `sc_ufix` data-types.

6.6 Summary

This chapter has demonstrated the powerful capabilities that SystemC provides for creating parameterized modules and channels. These capabilities make it possible for designers to create modules and channels that can be widely reused, thus leveraging their design investment.

In the following chapters we will see examples of how such parameterized modules and channels can be applied within the system-level design flow.

7

Interface and Channel Design

7.1 Introduction

Interfaces and channels play a key role in SystemC designs. As discussed in chapter 2, interface classes are used to declare the access methods that channels provide. Channels are used to implement the access methods declared within interfaces. Together, interfaces and channels provide a powerful and very flexible way to model communication and synchronization within systems.

This chapter provides a variety of design guidelines and techniques that are useful when creating and using your own interface and channel classes.

7.2 Interface Design

By carefully designing and selecting the interface classes used in your design, you can reduce the overall modeling effort, make design refinement easier, and increase opportunities for design reuse. The following sections provide some basic guidelines for designing interface classes.

Minimize the number of distinct interfaces.

Interfaces allow you to separate the declaration of the methods implemented within a channel from the actual implementation of those methods within a channel. This separation makes it possible to easily substitute one channel for another within SystemC as long as they implement the same interface. The fact that this can be easily accomplished in SystemC provides powerful design exploration and refinement capabilities. However, if channels unnecessarily use *different* interfaces when a common interface would suffice, the ease of substitution is lost.

Here's a simple example. Within the SystemC distribution there is the `sc_fifo` primitive channel that implements simple read (`sc_fifo_in_if<T>`) and write (`sc_fifo_out_if<T>`) interfaces. These interfaces don't provide any functions for specifying the size of the FIFO. Instead, the size of the FIFO is specified and fixed at the time of `sc_fifo`'s construction. If any write requests are ever made to `sc_fifo` when the FIFO is full, the write requests will block, so that data sent through the FIFO is always reliably delivered.

However, alternate FIFO implementations also might be desirable in certain cases. For example, it might be useful to implement a new FIFO `sc_fifo_discard` that discards items written to it once it is full rather than blocking the write request. Or it might be useful to have a `sc_fifo_stats` class that gathers statistics about the number of items within the FIFO. In both cases, it is desirable to reuse the existing `sc_fifo_in_if<T>` and `sc_fifo_out_if<T>` interface classes for these new FIFOs. This allows the FIFO implementations to be easily swapped during the process of design exploration and refinement simply by instantiating the new FIFO channel in place of the old. Port instances that access the channel do not need to be modified at all since they continue to use the same FIFO interfaces.

Layer specialized interfaces on more general interfaces and use the more general interfaces as much as possible to increase opportunities for channel reuse.

To increase the chances that channels can be reused, it is often useful to separate interface classes into layers. More specialized interfaces then inherit from the more general interfaces. Modules can then use the least-specialized interface that will provide the needed functionality, but channels that implement more-specialized interfaces in addition to the less-specialized ones can still be attached to such modules.

Here's a simple example of this. Assume we are modeling a bus channel that implements simple read and write transactions. We can declare the interface for these transactions as:

```
class simple_rw_if : virtual public sc_interface
{
public:
    virtual void read (unsigned addr, char* data) = 0;
    virtual void write(unsigned addr, char* data) = 0;
};
```

Let's assume that in some cases we want to model bus channels that support burst read and write transactions. Since all of the busses that support burst transactions will also support the simple nonburst transactions, we can layer this more specialized interface on the more general interface:

```
class burst_rw_if : public simple_rw_if
{
public:
    virtual void burst_read
      (unsigned addr, char* data, unsigned n) = 0;
    virtual void burst_write
      (unsigned addr, char* data, unsigned n) = 0;
};
```

Now let's assume we are modeling a simple module that does not need to use burst transactions. We can write:

```
class simple_module : public sc_module
{
public:
    sc_port<simple_rw_if> rw_port;
};
```

The port `rw_port` of this module can now be attached to both kinds of bus channels—those that implement only the `simple_rw_if` interface, as well as those that implement the `burst_rw_if` interface. In the case where `rw_port` is attached to a channel implementing the `burst_rw_if` interface, the rules of C++ insure that the burst transaction functions cannot be accessed.

Use class inheritance to group common interface methods and to reduce code duplication.

It can sometimes be advantageous to create interface classes even if you expect that they will never be directly used by modules or channels, but will only serve as base classes for other interface classes. This is particularly true if the base interface classes can serve as a common base class for several other interface classes.

Let's look again at the `simple_rw_if` interface class within the example above. Assume that `simple_rw_if` will never be used directly by modules or channels, but that we have another interface class such as `complex_rw_if` that also declares `read` and `write` methods as in `simple_rw_if`. In this case we can make `simple_rw_if` a common base class of both `burst_rw_if` and complex_rw_if. The advantage is that the overall code will be more compact and maintainable because the read and write methods are only declared once.

Create a unified interface class from separate interface classes using C++ multiple inheritance.

Occasionally you may find that you have several separate interface classes that you wish to combine into a new interface class, so that all of the methods within the separate interfaces can be accessed via the single unified interface. C++ provides a simple and elegant way to achieve this using multiple inheritance—simply define the unified interface to inherit from each of the separate interface classes, and don't add any members within the unified interface. The code will look similar to:

```
class unified_if
  : public first_if,
  : public second_if
{};
```

Or, if the interface classes have template arguments, the code will be similar to:

```
template <T>
class unified_if
  : public first_if<T>,
  : public second_if<T>
{};
```

We will also see an example of this technique used within chapter 8 with the simple_bus_unified_master_if interface class.

7.3 Primitive versus Hierarchical Channels

In chapter 2 we introduced the concept of channels in SystemC and noted that there are two separate kinds of channels that are supported: primitive channels (classes derived from sc_prim_channel) and hierarchical channels (classes derived from sc_channel). Now we will point out some of the key differences between these two channel types so that you can select the right type when creating a new channel.

Assume that you need to create a channel and you have a particular channel implementation in mind. In this case the rules to decide whether to use a primitive or hierarchical channel are straightforward:

1. You must use a primitive channel if you need to use the `request_update/update` scheme within your channel implementation (section 2.4.3).

2. You must use a hierarchical channel if your channel will contain processes, ports, modules, or other channels.

3. If you still have a choice, choose the primitive channel since it should generally use less memory.

If it seems that your channel implementation needs to use the `request_update/update` scheme and also contains processes, ports, modules, or other channels, then you need to redesign your channel since SystemC does not support this. In this case, the solution may be to split the channel into two channels—a primitive one and a hierarchical one. Let the primitive channel hold the data that needs to be updated. It is quite likely that there is already a channel available (such as `sc_signal<T>`) that you can use for this purpose. Then instantiate the primitive channel in the to-be-created hierarchical channel. See the `hw_fifo_wrapper` hierarchical channel presented later in this chapter for a channel implementation along these lines.

In general, creating new primitive channels that utilize the `request_update/update` scheme requires a fair amount of expertise. Most SystemC users will instead create hierarchical channels or primitive channels that do not use `request_update/update`. However, when simulation performance is critical, it is sometimes possible to create a new primitive channel that uses `request_update/update` to replace a hierarchical channel that may contain a number of processes and/or embedded channels or modules. Such substitutions can dramatically increase simulation performance by reducing the total number of process instances within the design and the amount of context switching between processes during simulation. However, users should generally only undertake such substitutions if there is evidence that the number of processes and/or context switches within the original hierarchical channel is large and can be reduced. If the new primitive channel and the old hierarchical channel use the same interface, such substitutions can be made with minimal impact on the code for the design since none of the attached modules need to be modified.

It is possible to perform *communication refinement* (see chapter 9) with both primitive channels and hierarchical channels. During the process of refinement, interfaces play a key role by allowing one channel to easily be swapped with another. But channel *implementations* are not usually refined—instead one channel is generally substituted with another. Thus a primitive channel could be swapped with a hierarchical channel or vice-versa.

7.4 Primitive Channel Examples

In this section we will show some examples of primitive channels and demonstrate some of their important modeling capabilities. In particular we will show how primitive channels aid in developing deterministic models of concurrent systems.

The most important capability provided by primitive channels is the ability to model delta-cycle communication delays via the `request_update/update` scheme. Using these capabilities, user defined channels that propagate state changes with infinitesimally small delays can be constructed. This makes it possible to construct designs that model concurrent and distributed systems despite the fact that the SystemC design itself will be simulated on a sequential processor where only a single process is executing at a given point. A typical example of a primitive channel that uses the `request_update/update` scheme is `sc_signal<T>`, introduced in chapters 2 and 4. As mentioned before, the behavior of `sc_signal<T>` is similar to the VHDL signal in that a new value assigned to a signal does not take effect immediately—instead there is always a delta-cycle delay before the new value takes effect. (This makes it possible, for example, to use two `sc_signal` objects to model two registers that swap values on a clock edge. In comparison, two software variables cannot swap values without the introduction of a third temporary variable.) Also recall that `sc_signal` does not allow more than one process to assign values to it—this instead is handled by `sc_signal_resolved` and `sc_signal_rv`, described later.

A simplified version of the interface that supports reading from an `sc_signal<T>` channel is:

```
template <class T> class sc_signal_in_if
  : virtual public sc_interface
{
public:
    // get the value changed event
    virtual const sc_event& value_changed_event() const = 0;
    // read the current value
    virtual const T& read() const = 0;
};
```

A simplified version of the interface that supports writing to (and reading from) a signal is:

```
template <class T> class sc_signal_inout_if
  : public sc_signal_in_if<T>
{
public:
    // write the new value
    virtual void write( const T& ) = 0;
};
```

And finally, a simplified version of the sc_signal<T> primitive channel itself is:

```
template <class T> class sc_signal
  : public sc_signal_inout_if<T>,
    public sc_prim_channel
{
public:
    // get the default event
    virtual const sc_event& default_event() const
    { return m_value_changed_event; }

    virtual const sc_event& value_changed_event() const
    { return m_value_changed_event; }

    virtual const T& read() const
    { return m_cur_val; }

    virtual void write( const T& value_)
    {
        m_new_val = value_;
        if( !( m_new_val == m_cur_val ) )
            request_update();
    }

protected:
    virtual void update()
    {
        if ( !( m_new_val == m_cur_val )) {
            m_cur_val = m_new_val;
            m_value_changed_event.notify(SC_ZERO_TIME);
        }
    }

    T              m_cur_val;
    T              m_new_val;
    sc_event       m_value_changed_event;
};
```

Let's walk through the scenario where one process writes a new value to the signal and causes another process to read the newly written value.

The writing process will call the `write(const T&)` method of the channel, which in turn will store the new value in the `m_new_val` data member, and then will call `request_update()` if the new value is different than the old value. The `request_update()` function will inform the SystemC scheduler that the `update()` method for this `sc_signal` instance should be called in the next simulation phase (i.e., during the update phase of the current delta-cycle.). The `write()` method then immediately returns before the new value has been propagated to the reading process, and the writing process and other processes in the design are free to perform additional `write()` operations to other signals. Once the SystemC scheduler advances to the update phase of the simulation cycle, the `update()` method of every primitive channel that called `request_update()` in the prior simulation phase is now invoked, causing the new value to be assigned to the current value and causing the `m_value_changed_event.notify()` call if the new value is different from the current. This event notification will cause the reading process to be resumed in the next simulation phase (since the reading process will have been waiting to resume execution on this event). Once the reading process is resumed, it will call the `read()` function to actually obtain the new value.

The above discussion illustrates how the simple and general delta-cycle scheduling capabilities provided by `request_update/update`, along with the event notification capabilities provided by `sc_event`, can be used to construct specific primitive channels in which communication is delayed by a delta-cycle. As we shall see shortly, delta-cycle communication delays are also useful for other types of primitive channels.

A second common modeling use of primitive channels is for resolution or arbitration of simultaneous actions within a design. (Within SystemC, two actions are said to be *simultaneous* if they occur within the same simulation phase and either is free to execute before the other.) Resolution or arbitration is often needed when multiple processes try to change the state of a particular channel simultaneously.

An example of a primitive channel that performs resolution is `sc_signal_resolved`, which is used to model a one-bit wide signal that may have multiple processes driving it. This signal may be driven with the values `0`, `1`, `X`, or `Z`, and the channel implementation must examine all of the driving values to determine the resolved value that the signal will take. (There is also a channel `sc_signal_rv<W>` that is similar to `sc_signal_resolved` but which supports signals wider than one bit.)

The implementation of sc_signal_resolved is quite similar to the implementation of sc_signal shown above. In fact, sc_signal_resolved inherits directly from sc_signal<sc_logic> and basically just overrides the write() and update() methods. Within the write() method it stores the current driving value of each process that assigns to the signal, and then calls request_update(). Then within the update() method a resolution function is called to compute a new value for the signal based on all of the current driving values.

Note that sc_signal_resolved gathers all of its incoming state change requests during the evaluation phase of the simulation cycle, and then deterministically resolves the competing requests into a single next state that is made available to the rest of the design after the completion of the update phase. This approach of gathering state change requests during the evaluation phase, determining the next state of the channel during the update phase, and then propagating the new channel state to the rest of the design during the next simulation phase is the key idea behind designing primitive channels that use request_update/update. Consider that a similar approach could be used to model a bus channel in which several devices might request ownership of the bus simultaneously—in this case the bus channel would gather the bus ownership requests during the evaluate phase and determine the next bus master during the update phase based on the requesting device priorities.

Now let us briefly consider how to model a FIFO in SystemC. Similar to the sc_fifo<T> provided in the SystemC distribution, let's assume we wish to provide a blocking write interface:

```
template <class T> class sc_fifo_out_if
  : virtual public sc_interface
{
public:
    // blocking write
    virtual void write( const T& ) = 0;

    // other methods not shown
};
```

and also a blocking read interface:

```
template <class T> class sc_fifo_in_if
  : virtual public sc_interface
{
public:
    // blocking read
```

```
    virtual void read( T& ) = 0;

    // query the number of items in FIFO
    virtual int num_available() const = 0;

    // other methods not shown
};
```

Like sc_fifo<T>, let's allow only a single process to write to the FIFO and a single process to read from it. A simple channel implementation of a FIFO would inherit from both of the above interfaces and provide implementations for the write and read methods that directly write and read the data to/from an array indexed as a circular buffer, suspending write operations if there is no space available, and suspending read operations if there is no data available. (A working example of such a simple FIFO implementation is provided within the *examples/systemc/simple_fifo* directory within the SystemC distribution.)

Now consider the following scenario with this FIFO implementation. Assume that the writing process and the reading process are triggered by the same event (e.g., a rising clock edge) so that they execute in the same simulation phase. Assume that the FIFO is empty and the writer writes one item into it, and that the reader calls the num_available() method. What number of items will the reader see?—either zero or one in this case, since the order of process execution within any simulation phase is unspecified. It is conceivable that this unpredictable behavior could introduce subtle problems into the rest of the design. Because of these sorts of issues, the actual implementation of sc_fifo<T> provided with SystemC relies on the request_update/update scheme. In the actual implementation, all writes to the FIFO do not change the externally visible state of the FIFO until the next delta-cycle, thus eliminating any unpredictable behavior.

It is important to realize that not all primitive channels use the request_update/update scheme. An example of a channel that does not use this scheme is sc_mutex, presented in simplified form below:

```
class sc_mutex_if : virtual public sc_interface
{
public:
    // blocks until mutex could be locked
    virtual void lock() = 0;

    // returns false if mutex could not be locked
    virtual bool trylock() = 0;
```

```
    // unlocks mutex
    virtual void unlock() = 0;
};


class sc_mutex : public sc_mutex_if,
                 public sc_prim_channel
{
public:
    virtual void lock() {
        while (_locked)
            wait(_free);

        _locked = true;
    }

    virtual bool trylock() {
        if (_locked) return false;

        _locked = true;
        return true;
    }

    virtual void unlock() {
        _locked = false;
        notify(_free);
    }

protected:
    sc_event _free;
    bool     _locked;
};
```

In this implementation of `sc_mutex`, if a number of processes are waiting to lock the mutex and the mutex is then unlocked, all of the waiting processes will be resumed via the `notify(_free)` call. However, only the process that happens to execute first after the `unlock` call will succeed in locking the mutex. All of the other contending processes will wake up, see that the mutex is locked again, and then resume their wait for the mutex.

Does SystemC or `sc_mutex` provide any guarantees about *which* process will succeed in locking the mutex when there are several contenders? The answer is *no*. The reason is that we wish to avoid over-specifying systems. If SystemC explicitly specified the order of process execution within a simulation phase, or if `sc_mutex` gathered all of its lock requests during the evaluate phase and then somehow deterministically selected one during the update phase, then such ordering constraints would need to be maintained as the design was refined to implementation. Maintaining such arbitrary constraints during the refinement flow would be very difficult and might lead to nonoptimal design implementations. So here we have an example of a primitive channel that *could* be written to gather state change requests during the evaluate phase and to arbitrate them during the update phase, but we see that this is probably not a good design approach in this case.

As an exercise, you might consider how to implement a different mutex primitive channel that is more efficient in the case where a large number of processes are waiting to lock the mutex. This implementation would avoid resuming all of the waiting processes except the one securing the lock. A tip: allocate a separate `sc_event` object within the mutex for each process that tries to lock the mutex. A further refinement of this mutex implementation could call `rand()` to insure that processes are selected for resumption in a truly random order.

We say that a SystemC design is *deterministic* if the design will always produce the same result on any compliant SystemC simulator when a given stimulus is applied. Determinism is often a desirable property of designs because:

1. Given the same stimulus, the same design will produce the same result even with different versions of SystemC simulators, on different machine platforms, or even if the same simulator executes processes in random or varying orders within a simulation phase.

2. A real-world implementation that has been properly derived from a deterministic design specification will always produce the same result for the same stimulus as the original design.

There are several approaches used to construct SystemC designs that will always be deterministic.

The first approach is based on the careful and consistent use of primitive channels that use `request_update/update`. To determine if a design adheres to this approach, review the design and mentally "flatten away" the design structure leaving only the processes and primitive channels within the design. (Modules, ports, and hierarchical channels are flattened away, and code implementing hierarchical channel methods is "merged" into the calling process.) Now determine if all communication between processes goes through primitive channels that:

- Use the `request_update/update` scheme, thus delaying all communication between processes by a delta-cycle (e.g., `sc_signal`, `sc_` signal_resolved, sc_signal_rv, sc_fifo, sc_buffer). Note that the design can use a combination of such primitive channel types—it does not need to use exclusively one type of primitive channel.

- If any primitive channel may simultaneously receive state change requests from multiple processes, the channel must deterministically arbitrate or resolve the requests in a manner that is not affected by the order in which the requests were received. (e.g., sc_signal_resolved, sc_signal_rv, but *not* sc_mutex and sc_semaphore.)

If the design meets these requirements, it will have the deterministic properties described above.

A second approach to achieving determinism within SystemC designs is to rely on an overall model of computation that is known to guarantee determinism. An example of such a model of computation is the dataflow MOC introduced in chapter 5. If such a model of computation is used consistently throughout a design, then design execution will be deterministic even if primitive channels that use request_update/update are not employed.

A third approach to achieving determinism within SystemC designs is to rely on a *two-phase synchronization scheme*, which is discussed in chapter 8.

SystemC designs that do not follow one of the above approaches (or which use some combination of the above approaches) may still be deterministic, but a more in depth analysis may be needed to determine that this is the case.

7.5 Hierarchical Channel Example

Hierarchical channels were introduced in chapter 2, where we indicated that hierarchical channels (classes derived from sc_channel) are modules that implement one or more interfaces. Like modules, hierarchical channels may have embedded child modules, channels, or processes. And similar to primitive channels just discussed, hierarchical channels provide implementations for the methods declared in one or more interface classes. Because hierarchical channels can be used to encapsulate both the structural elements of a design as well as the communication protocols or methods that are used in that part of the design, they can be used to model complex communication within a design in a much more elegant way than is possible in languages such as VHDL and Verilog.

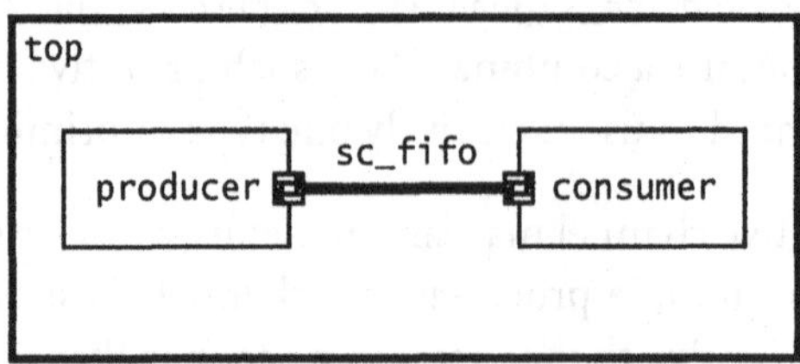

Figure 7.1: A producer and consumer module communicating via a primitive channel, sc_fifo

We will now present an example of a hierarchical channel named hw_fifo_wrapper<T> that implements the same interfaces as sc_fifo<T>. Before introducing hw_fifo_wrapper<T>, let us first present a simple, fully working producer/consumer design that writes and reads characters from an sc_fifo<char> primitive channel in the code that follows. We depict this design in figure 7.1.

```
#include <systemc.h>

class producer : public sc_module
{
public:
    sc_port<sc_fifo_out_if<char> > out;

    SC_HAS_PROCESS(producer);

    producer(sc_module_name name) : sc_module(name) {
        SC_THREAD(main);
    }

    void main() {
        const char *str = "Visit www.systemc.org!\n";
        const char *p = str;

        while (true) {
            if (rand() & 1) {
                out->write(*p++);
                if (!*p) p = str;
            }

            wait(1, SC_NS);
        }
    }
};
```

```
class consumer : public sc_module
{
public:
    sc_port<sc_fifo_in_if<char> > in;

    SC_HAS_PROCESS(consumer);

    consumer(sc_module_name name) : sc_module(name) {
        SC_THREAD(main);
    }

    void main() {
        char c;

        while (true) {
            if (rand() & 1) {
                in->read(c);
                cout << c;
            }

            wait(1, SC_NS);
        }
    }
};

class top : public sc_module
{
public:
    sc_fifo<char> fifo_inst;
    producer prod_inst;
    consumer cons_inst;

    top(sc_module_name name, int size) :
        sc_module(name),
        fifo_inst("Fifo1", size),
        prod_inst("Producer1"),
        cons_inst("Consumer1")
    {
        prod_inst.out(fifo_inst);
        cons_inst.in(fifo_inst);
    }
};

int sc_main (int argc, char *argv[])
{
    int size = 10;

    top top1("Top1", size);
```

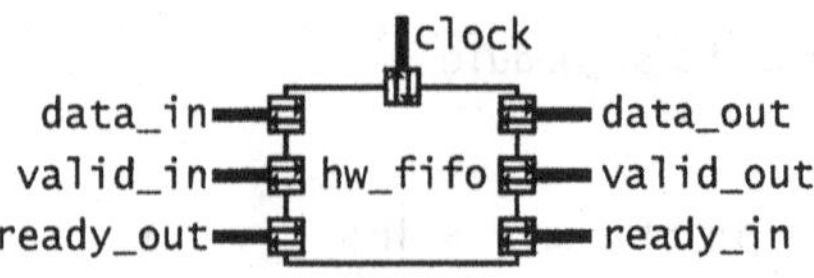

Figure 7.2: Schematic symbol for hw_fifo

```
    sc_start(1000, SC_NS);
    cout << endl << endl;
    return 0;
}
```

In the above design once a nanosecond the producer will write one character to the FIFO instance with 50 percent probability, while the consumer will read one character from the FIFO with 50 percent probability. Because of the blocking nature of sc_fifo<T> read and write operations, all data is reliably delivered through the FIFO despite the varying rates of production and consumption.

Now let's present a model of a clocked RTL hardware FIFO named hw_fifo<T> and assume we wish to use this model in place of the sc_fifo<T> instance above. This model uses a signal-level ready/valid handshake protocol for both the FIFO input and output. It uses a circular buffer implemented within a dynamically allocated _data array to store and retrieve the items within the FIFO. The code for the hw_fifo<T> module follows and its schematic symbol is depicted in figure 7.2.

```
template <class T> class hw_fifo : public sc_module
{
public:
    sc_in<bool>  clk;

    sc_in<T>     data_in;
    sc_in<bool>  valid_in;
    sc_out<bool> ready_out;

    sc_out<T>    data_out;
    sc_out<bool> valid_out;
    sc_in<bool>  ready_in;

    SC_HAS_PROCESS(hw_fifo);

    hw_fifo(sc_module_name name, unsigned size)
      : sc_module(name), _size(size)
```

```
    {
        assert(size > 0);
        _first = _items = 0;
        _data = new T[_size];

        SC_METHOD(main);
        sensitive << clk.pos();

        ready_out.initialize(true);
        valid_out.initialize(false);
    }

    ~hw_fifo() { delete[] _data; }

protected:

    void main()
    {
        if (valid_in.read() && ready_out.read())
        {
            // store new data item into fifo
            _data[(_first + _items) % _size] = data_in;
            ++_items;
        }

        if (ready_in.read() && valid_out.read())
        {
            // discard data item that was just read from fifo
            -- _items;
            _first = (_first + 1) % _size;
        }

        // update all output signals
        ready_out = (_items < _size);
        valid_out = (_items > 0);
        data_out = _data[_first];
    }

    unsigned _size, _first, _items;
    T* _data;
};
```

It should be clear that we cannot use `hw_fifo` directly in place of `sc_fifo`, since the latter *implements* the `sc_fifo_in_if` and `sc_fifo_out_if` interfaces, while the former doesn't implement any interfaces at all, but has ports that connect to signals (or, to be more precise, has ports that *use* the `sc_signal_in_if` and `sc_signal_out_if` interfaces).

To solve this problem, we now define a hierarchical channel hw_fifo_wrapper<T> that implements the sc_fifo_in_if<T> and sc_fifo_out_if<T> interfaces, and which contains an instance of hw_fifo<T>. In addition hw_fifo_wrapper<T> contains sc_signal instances that are used to interface with hw_fifo<T>. Since hw_fifo<T> has a clock port, we also need a clock port on hw_fifo_wrapper<T> to feed in the clock signal to the hw_fifo<T> instance.

The key functionality within hw_fifo_wrapper<T> is found within the implementations of its read() and write() methods. We see that these methods implement the required signal-level ready/valid handshake protocol with hw_fifo<T> whenever a read or write transaction occurs. Note in particular that the protocol will properly suspend read() or write() transactions if hw_fifo<T> is not ready to complete the operation.

```
template <class T> class hw_fifo_wrapper
  : public sc_module,
    public sc_fifo_in_if<T>,
    public sc_fifo_out_if<T>
{
public:
    sc_in<bool> clk;

protected:
    // embedded channels

    sc_signal<T>    write_data;
    sc_signal<bool> write_valid;
    sc_signal<bool> write_ready;

    sc_signal<T>    read_data;
    sc_signal<bool> read_valid;
    sc_signal<bool> read_ready;

    // embedded module

    hw_fifo<T> hw_fifo;

public:

    hw_fifo_wrapper(sc_module_name name, unsigned size)
      : sc_module(name), hw_fifo("hw_fifo1", size)
    {
        hw_fifo.clk(clk);

        hw_fifo.data_in  (write_data);
        hw_fifo.valid_in (write_valid);
```

```
        hw_fifo.ready_out(write_ready);

        hw_fifo.data_out (read_data);
        hw_fifo.valid_out(read_valid);
        hw_fifo.ready_in (read_ready);
    }

    virtual void write(const T& data)
    {
        write_data = data;
        write_valid = true;

        do {
            wait(clk->posedge_event());
        } while (write_ready != true);

        write_valid = false;
    }

    virtual T read()
    {
        read_ready = true;

        do {
          wait(clk->posedge_event());
        } while (read_valid != true);

        read_ready = false;
        return read_data.read();
    }

    virtual void read(T& d) { d = read(); }

    // Provide dummy implementations for unneeded
    // sc_fifo<T> interface methods:

    virtual bool nb_read(T&)
      { assert(0); return false; }
    virtual bool nb_write(const T&)
      { assert(0); return false; }
    virtual int num_available() const
      { assert(0); return 0; }
    virtual int num_free() const
      { assert(0); return 0; }
};
```

Now that we've defined hw_fifo_wrapper<T>, we can use it to replace the sc_fifo<T> instance within our original design. All we need to do

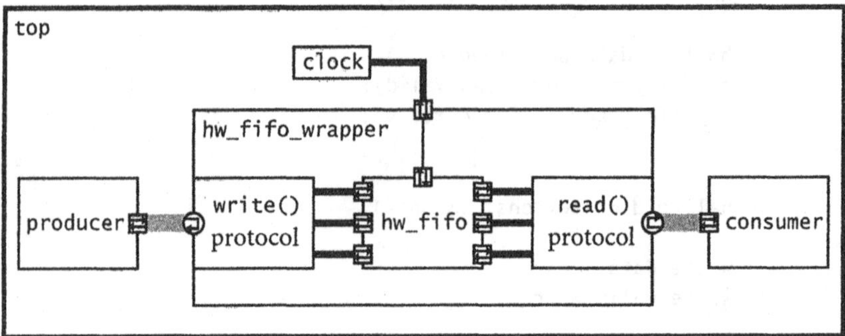

Figure 7.3: The producer and consumer modules now communicating via the hw_fifo_wrapper hierarchical channel.

is to add a hardware clock to drive the additional clock port that is now on hw_fifo_wrapper<T>. Below we show the minor changes that need to be made to the original design code in order to incorporate hw_fifo_wrapper<T>, and we depict in figure 7.3 the complete design now using hw_fifo_wrapper<T>.

```
class top : public sc_module
{
public:
    hw_fifo_wrapper<char> fifo_inst; // changed
    producer prod_inst;
    consumer cons_inst;
    sc_clock clk;                    // added

    top(sc_module_name name, int size) :
        sc_module(name),
        fifo_inst("Fifo1", size),
        prod_inst("Producer1"),
        cons_inst("Consumer1"),
        clk("c1", 1, SC_NS)          // added
    {
        prod_inst.out(fifo_inst);
        cons_inst.in(fifo_inst);
        fifo_inst.clk(clk);          // added
    }
};
```

We can now simulate the design again, this time with hw_fifo<T> providing the FIFO functionality instead of sc_fifo<T>. Simulation will show that the overall design behavior hasn't changed (though some of the timing may now be different), and in particular that data now is reliably delivered through `hw_fifo<T>` despite the varying production and consumption rates.

This example illustrates how designers can create hierarchical channels that encapsulate both design structure and communication protocols and easily swap channels in order to perform design exploration or refinement. It also shows how the reuse of interface classes facilitates channel swapping.

In chapter 9 we will see that hierarchical channels can play an important role in the process of communication refinement. We will also see that hierarchical channel methods that implement protocols (such as the `hw_fifo_wrapper::read()` method above) are closely related to the concept of adapters used during communication refinement.

7.6 Summary

In this chapter we've explored some of the major aspects of interface and channel design. We've noted the differences between primitive and hierarchical channels and have presented guidelines for selecting between the two. We've discussed some of the important primitive channels that are included within the standard SystemC elementary channels, and we've shown how primitive channels can be used to create deterministic models of systems. We've also presented a hierarchical channel example and have seen how hierarchical channels can be used to encapsulate other channels, modules, processes, and protocol implementations.

In the next chapter we will present a transaction level model of a design and see how channels play a key role in modeling the design. We will also see some additional channel design techniques that SystemC offers.

8

Transaction-Level Modeling

8.1 Introduction

Abstraction is a powerful technique for the design and implementation of complex systems. It allows us to tackle complexity by first hiding unnecessary details and then working them out later. Different amounts of details correspond to different levels of abstraction. In chapter 4 we have seen the register-transfer and behavioral levels of abstraction. These levels hide details such as gates and latency of computation below the clock-cycle level. But structurally these levels are *pin-accurate*; that is, the exact connections and registers are explicit at the structural boundaries, thereby placing certain restrictions on the way data goes in and out of modules.

In this chapter we introduce a modeling style called *transaction-level modeling* (TLM). Transaction-level modeling is a high-level approach to modeling digital systems where details of communication among modules are separated from the details of the implementation of the functional units or of the communication architecture. Communication mechanisms such as busses or FIFOs are modeled as channels, and are presented to modules using SystemC interface classes. Transaction requests take place by calling interface functions of these channel models, which encapsulate low-level details of the information exchange. In other words, at the transaction-level, the emphasis is more on the *functionality* of the data transfers—what data are transferred to and from what locations—and less on their actual *implementation* (that is, on the actual protocol used for data transfer). This approach makes it easier for the system-level designer to experiment, for example, with different bus architectures (all supporting a common abstract interface) without having to recode models that interact with any of the busses, provided these models interact with the bus through the common interface.

From the point of view of the clients of the communication channels, it is also much easier to use. Recall that in behavioral-level modeling, even though we did not have to worry about the exact clock-cycle latency of operations, we still needed to treat handshake signals explicitly, and we still needed to alter the handshaking sequences if the communication protocol changed. In contrast, in transaction-level modeling, such synchronization details are typically abstracted into the categories of *blocking* and *nonblocking I/O*, and, in the case of busses, priorities may be assigned to the clients, and arbitration can be modeled in a centralized way.

Transaction-level modeling also enables higher simulation speed than pin-based interfaces, through the suppression of "uninteresting" details. For instance, in the real world a large burst-mode transfer may take many actual clock cycles to complete. In most of these clock cycles the bus is merely doing routine work and those clients that have pending bus requests are just waiting. If we view the burst-mode transfer as a single operation, there is no need to devote simulation time to these "uninteresting" clock cycles. Depending on whether the model needs to be bus-cycle-accurate or not, different strategies can be applied to take advantage of this, resulting in significant savings in simulation time. As we will see in the next section, even when a transaction-level model needs to be cycle-accurate, it still may simulate much faster than a typical cycle-accurate RTL model.

Transaction-level models can be used anytime their increased level of abstraction is beneficial. Typically transaction-level models are used for functional modeling (both timed and untimed), platform modeling, and for constructing testbenches.

An example of an untimed functional transaction-level model that we have already discussed is `sc_fifo`. The `read` and `write` methods of this channel represent its transaction interfaces. The `sc_fifo` channel models the functionality of a typical FIFO, but the way in which the FIFO is modeled within `sc_fifo` is likely to be very different, and much simpler, than an actual hardware implementation of a FIFO. Even an actual software implementation of a FIFO on top of a preemptive RTOS will likely be more complicated than `sc_fifo` because of the need to insure that data accessed by multiple preemptive threads is handled safely.

To illustrate some of the transaction-level modeling concepts applied to a bus model, we present the following highly simplified model of a bus that supports burst read and burst write transactions:

```
class very_simple_bus_if : virtual public sc_interface
{
public:
    virtual void burst_read (char *data,
                             unsigned addr,
                             unsigned length) = 0;
    virtual void burst_write(char *data,
                             unsigned addr,
                             unsigned length) = 0;
};

class very_simple_bus
  : public very_simple_bus_if,
    public sc_channel
{
public:
    very_simple_bus(sc_module_name nm, unsigned mem_size,
        sc_time cycle_time) : sc_channel(nm),
                              _cycle_time(cycle_time)
    {
        // we model bus memory accesses using an embedded
        // memory array
        _mem = new char [mem_size];

        // set initial value of memory to zero
        memset(_mem, 0, mem_size);
    }

    ~very_simple_bus() { delete [] _mem; }

    virtual void burst_read(char *data, unsigned addr,
                            unsigned length)
    {
        // model bus contention using a mutex,
        // but no arbitration rules
        _bus_mutex.lock();

        // block the caller for length of burst transaction
        wait(length * _cycle_time);

        //  copy the data from memory to requester
        memcpy(data, _mem + addr, length);

        // unlock the mutex to allow others access to the bus
        _bus_mutex.unlock();
    }

    virtual void burst_write(char *data, unsigned addr,
                             unsigned length)
```

```
    {
        _bus_mutex.lock();

        wait(length * _cycle_time);

        // copy the data from requestor to the memory
        memcpy(_mem + addr, data, length);

        _bus_mutex.unlock();
    }

protected:
    char* _mem;
    sc_time _cycle_time;
    sc_mutex _bus_mutex;
};
```

This very simple bus does not model a number of important aspects of a real-world bus such as bus-request arbitration, interrupted burst transactions, memory wait states, and bus masters that need to continue to be active while they have uncompleted bus requests. Also, to keep the model simple, we have modeled the memory that the bus can access as an array within the bus rather than external to the bus. We will see in our next example how we can use a very similar modeling style to accurately model the activity on the bus on each clock cycle, thus enabling more detailed effects such as interrupted burst transactions to be very accurately modeled. It is important to understand that aspects such as contention, arbitration, interrupts, and cycle-accuracy can be accurately modeled with TLM models without resorting to pin-accurate models.

In the following sections we will illustrate transaction-level modeling by way of a simple bus example, fully exploiting the SystemC notions of interfaces and channels. TLM is by no means limited to busses and FIFOs; the same principles may be applied to other types of higher-order communication mechanisms.

8.2 The Simple Bus Design

We will now present the `simple_bus` design and discuss the modeling techniques used within it. The `simple_bus` design is an example of a high performance, cycle-accurate, platform transaction-level model in SystemC.

To fully understand the `simple_bus` discussion within this chapter, you will need to download and inspect the design source code. You may want to do this right now if you don't already have the source code. The complete code and detailed documentation for simple_bus can be downloaded from *www.systemc.org*. (In SystemC 2.0, simple_bus is a separate download, but in later versions of SystemC it should be included within the examples directory.) The amount of code for simple_bus is about one thousand lines.

The following discussion introduces simple_bus and focuses on techniques used within it to create compact and high performance models. More detailed documentation about simple_bus is also available with the software distribution.

Why are transaction-level models similar to simple_bus important? We believe that such models address a critical need in system design today by providing executable system models that are:

- Relatively easy to develop, understand, use, and extend
- Able to accurately model both the hardware and software components of a system
- Capable of being constructed very early in the system design process, thus enabling designers to explore implementation alternatives and make design trade-offs before it's too late or too expensive to do so
- Fast and accurate enough to validate software before more detailed hardware models or implementations are available

A key point is that such models need to have very high simulation speed in order to allow meaningful amounts of software to be executed along with the hardware model. In addition, the models need to be capable of being fully cycle-accurate so that both hardware and software design teams have confidence in them. This allows the model to serve as an "agreed-upon contract" between the hardware and software teams before each team begins their detailed implementation. What's our definition of "very high simulation speed"? A rough number might be one hundred thousand clock cycles per second. With this speed, a model of a system might execute roughly a thousand times slower than its real-world implementation, making it possible to apply a variety of realistic tests that include the software parts of the system. As we will see later, the simple_bus design meets this performance metric while also offering cycle-accurate modeling.

8.3 Structure of the Simple Bus Design

The `simple_bus` design contains the following types of blocks:

- *Masters:* Examples of possible bus master blocks include CPUs and DSPs.
- *Bus:* The bus connects the masters and slaves and allows them to communicate using bus transactions.
- *Slaves:* Examples of possible slave blocks include ROMs, RAMs, I/O devices, and ASICs (e.g., hardware accelerators).
- *Arbiter:* The arbiter is used by the bus to select a request to execute from the competing bus requests.
- *Clock Generator:* The clock generator block provides a clock signal to most of the blocks in the design.

Masters can *initiate* transactions on the bus. Slaves merely respond to bus requests that they receive. When a master has been granted access to the bus, requests from other masters are queued by the bus and executed in later cycles.

Figure 8.1 shows the block diagram for the example `simple_bus`. In this design there are three masters and two slaves. The first master is a module that uses the blocking master interface presented below to access the bus. This blocking master module is an example of a high-level software model that generates transactions on the bus.

The second master is a module that uses the nonblocking master interface presented below. This nonblocking master module is similar to a detailed processor model (e.g., an instruction-set simulator model) that must still execute on every clock edge even if it is waiting for its bus transactions to complete.

The third master is a module that uses the direct master interface of the bus (presented below) to print debug information about the contents of the memories as the design executes. This direct master module is a bus monitor only (for use in the testbench) and does not represent a block that will exist in the real-world implementation.

The designer assigns a unique priority to each master that is used both to identify the masters to the bus and also to prioritize competing bus requests from different masters.

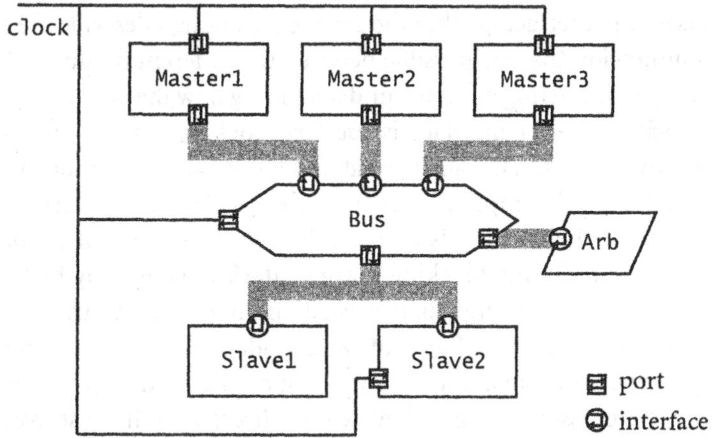

Figure 8.1: Block diagram for `simple_bus`

The first slave is a fast memory supporting single-cycle read/write operations. The second slave is a slow memory that takes a configurable number of cycles to complete a read/write operation. We refer to the additional cycles that a slave takes to complete an operation as *wait states* because all other activity on the bus waits until the operation completes.

Most modules (including the bus itself) are connected to the global clock signal. The only modules that are not connected to the clock are the fast memory and the arbiter. These two blocks do not need to be connected to a clock signal because they do not have any clocked processes within them. (The reason for this will be explained in section 8.8, "Common Questions" later in this chapter.)

Note that it is easy to add different kinds and numbers of masters or slaves to the simple bus design, and it is also easy to change the bus arbitration policies by replacing the arbiter module. It is easy to add new masters to the design since the masters connect to the bus using just one simple port connection. It is easy to add additional slaves to the design since the SystemC *multiport* feature is used (see section 8.8). And it is easy to change the arbiter in the design since the arbiter is a separate module from the bus.

8.4 Transaction Interfaces in Simple Bus

The transaction interfaces within the simple_bus design describe the types of communication that are possible between the different modules. These interfaces are the starting point for understanding how the design operates.

The blocking master interface is *used* by blocking masters and is *implemented* by the bus. The burst_read and burst_write methods within this interface model complete burst transactions. We use the term *blocking* here because these methods return only after the burst transaction has completed. Typically the blocking master interface is used by high-level software models that generate burst_read and burst_write transactions on the bus as they execute. (Such software models are not cross-compiled to a target processor and are not running on an instruction-set simulator of the target processor—instead they execute directly on the host workstation.) If only a single item needs to be read or written, the master simply uses the burst_read and burst_write methods with the length specified as one. (We will explain the use of the lock flag in section 8.6, "Overall Execution Scheme" of this chapter.)

```
class simple_bus_blocking_if
  : public virtual sc_interface
{
public:
    virtual simple_bus_status burst_read(
                              unsigned int unique_priority,
                              int *data,
                              unsigned int start_address,
                              unsigned int length = 1,
                              bool lock = false) = 0;
    virtual simple_bus_status burst_write(
                              unsigned int unique_priority,
                              int *data,
                              unsigned int start_address,
                              unsigned int length = 1,
                              bool lock = false) = 0;
};
```

The nonblocking master interface is used by nonblocking masters and is implemented by the bus. The read and write methods within this interface cause a single item to be transferred. The methods themselves return immediately, but the read or write transaction will take at least one clock cycle to actually complete. If competing bus requests exist, it may take more than once cycle to complete. After calling the read or write methods, the nonblocking master can call the get_status method on subsequent clock cycles to see if the transaction has completed yet. The nonblocking master interface is commonly used by processor models (e.g., instruction-set simulators). Such models typically cannot be suspended while they have outstanding bus requests since they still need to be activated on every clock cycle.

```
class simple_bus_non_blocking_if
  : public virtual sc_interface
{
public:
    virtual void read (unsigned int unique_priority,
                       int *data,
                       unsigned int address,
                       bool lock = false) = 0;
    virtual void write(unsigned int unique_priority,
                       int *data,
                       unsigned int address,
                       bool lock = false) = 0;

    virtual simple_bus_status get_status(
                       unsigned int unique_priority) = 0;
};
```

The direct master interface is used by "direct master" modules (explanation to follow) and is implemented by the bus. The `direct_read` and `direct_write` methods within this interface provide instantaneous read/write access to the slaves. The accesses go through the bus so that the address map of the bus can be used for proper routing of the requests. When these operations are performed the SystemC scheduler will not intervene and simulated time will not advance, and no modeling of bus contention or arbitration will occur. The direct master interface is useful for creating simulation monitors for the design, and it is also useful during initial functional modeling of the design when extremely high simulation performance is desired. Also, since these operations execute within a single SystemC thread, they are sometimes useful when you are debugging problems in the design. A key use of the direct master interface is in enabling software debuggers running on top of ISS models of CPUs to work. The direct master interface allows such software debuggers to read and write to addresses on the bus without having to map addresses to the appropriate slave, and without requiring that simulation time advances for the operations to complete. Once initial functional modeling is completed, the direct master interface should not be used within modules that will be part of the design implementation, but it can be used by modules that are part of the testbench.

```
class simple_bus_direct_if
  : public virtual sc_interface
{
public:
    virtual bool direct_read(int *data, unsigned int address) = 0;
    virtual bool direct_write(int *data, unsigned int address) = 0;
};
```

The slave interface is *used* by the bus and is *implemented* by every slave. (Strictly speaking, each slave is thus a channel. We will explain the reason for this approach in section 8.8, "Common Questions" later in this chapter.) The bus uses the `read` and `write` methods to transfer a single data item to or from the slave. The bus uses the `start_address` and `end_address` functions to map requests to the appropriate slave (each slave occupies a separate part of the overall address space). The bus uses the `direct_read` and `direct_write` methods (inherited from `simple_bus_direct_if`) to complete direct-mode accesses.

```
class simple_bus_slave_if
  : public simple_bus_direct_if
{
public:
    virtual simple_bus_status read (int *data,
                                     unsigned int address) = 0;
    virtual simple_bus_status write(int *data,
                                     unsigned int address) = 0;

    virtual unsigned int start_address() const = 0;
    virtual unsigned int end_address() const = 0;
};
```

The arbiter interface is *used* by the bus and is *implemented* by the arbiter. It is used to arbitrate competing bus requests. On each cycle the bus passes all of its outstanding requests to the arbiter module, which selects one for execution based on its arbitration policy. Outstanding requests are passed to the arbiter as a vector and a pointer to the selected request is returned.

```
class simple_bus_arbiter_if
  : public virtual sc_interface
{
public:
    virtual simple_bus_request *
        arbitrate(const simple_bus_request_vec &requests) = 0;
};
```

In order to keep the simple_bus design simple, we do not model more complex bus transactions such as pipelined transactions, split transactions, master wait states, and interrupt request/acknowledge schemes. These additional bus features can be modeled as an extension of the simple_bus design, but we will leave this as an exercise for the reader.

Now that we have examined the interfaces used within the simple_bus design, we can see how they are each incorporated into the bus itself by examining the simple_bus class declaration:

```
class simple_bus
  : public simple_bus_direct_if,
    public simple_bus_non_blocking_if,
    public simple_bus_blocking_if,
    public sc_channel
{
public:
    // ports
    sc_in_clk clock;
    sc_port<simple_bus_arbiter_if> arbiter_port;
    sc_port<simple_bus_slave_if, 0> slave_port;
    // constructors, member functions, processes, etc.
};
```

We see that the bus inherits from every interface that it implements, and that it contains ports for those interfaces that it uses.

8.5 Modeling Techniques for High Performance

Before exploring further the operation of simple_bus, let's pause for a moment and focus on one of the key goals of our design: high simulation performance. We have already seen enough of the design in terms of its transaction interfaces to point out some significant differences from typical cycle-accurate RTL models:

- Rather than relying on signals to model communication, we rely on *transactions*. A transaction is modeled using one function call that transfers both control and all of the data within the transaction from one module to another. The model is not pin-accurate—all of the data within a transaction can thus be bundled and passed much more efficiently.

- The model uses high-level data-types (whenever possible built-in C++ data types) rather than low-level bit-vectors, four-value bits (i.e., 01XZ), etc., as are commonly used in HDLs.

- Pointers to data are passed between modules within transactions, enabling one module to very efficiently copy blocks of data to another. We see an example of this within `very_simple_bus` above (section 8.1), and this technique is also used in `simple_bus` in the `burst_read` and `burst_write` transactions between masters and the bus.

- The SystemC *dynamic sensitivity* feature is used to eliminate unnecessary activations of processes. We see an example of this in the `very_simple_bus` code above, where processes that access the bus are completely suspended until their bus requests complete. This technique is also used in `simple_bus` for `burst_read` and `burst_write` transactions. In typical cycle-accurate RTL modeling, models must execute on every clock edge even if no work is being done. With this use of dynamic sensitivity we achieve a significant performance gain over typical RTL models because processes accessing the bus now are only activated when they have real work to perform.

There are some additional modeling techniques within `simple_bus` used to achieve high simulation speeds. We introduce them now and provide more details about them in the discussion that follows:

- Some modules (`fast_mem`, `arbiter`) are modeled without any processes at all! We can do this because these particular modules do not really need their own process—they are so simple that they can just let the bus process do all of their work when the bus actually sends requests to them. We will explain this further in section 8.8, "Common Questions," within this chapter.

- Where possible, we use `SC_METHOD` processes instead of `SC_THREAD` processes (e.g., in `simple_bus` and `slow_mem`). `SC_METHOD` processes generally consume much less memory and execute more quickly than `SC_THREAD` processes. In the `simple_bus` design, the blocks that are likely to be activated most frequently (`bus`, `arbiter`, `fast_mem`, and `slow_mem`) either use `SC_METHOD` processes or have no processes at all. As a result, most process activations during simulation are of `SC_METHOD` processes rather than `SC_THREAD` processes.

- We try to make frequently activated processes do as little work as possible. An example of this is within the `slow_mem` module. It has a clocked `SC_METHOD` process that decrements a counter that indicates when its wait states have completed. This process doesn't do any other work. The work of initializing the counter at the beginning of a memory access and the work of performing the actual read or write of the data is performed instead by the bus process when a read or write access to `slow_mem` is initiated or completed. (Even though this work is done by the bus process when it accesses `slow_mem`, the *description* of the work to be done is still contained within the `slow_mem` channel in the form of a channel method of `slow_mem`. This allows a clean separation of functionality to be maintained between different modules.)

The combination of modeling techniques such as those outlined above leads to very high simulation performance within transaction-level models as compared to typical RTL models.

8.6 Overall Execution Scheme within Simple Bus

We can now outline the overall execution scheme for the `simple_bus` design.

On the rising edge of the clock, masters execute and may send requests to the bus where they are stored. The bus maintains a set of all outstanding requests from masters that have not yet completed (including unfinished requests from past cycles).

On the falling edge of the clock, the bus calls the arbiter (using a function call that returns immediately) to select a request for execution. The arbiter will select one request based on its arbitration rules. Next, the bus looks up the address of the selected request in the bus's address map to determine which slave to send the request to. Then the `read()/write()` method of the appropriate slave is invoked. The slave `read()/write()` method will return immediately and will indicate if the slave issued a wait state. If the slave issued a wait state, the bus will prevent any other requests from gaining access to the bus and instead will reissue the current request to the slave on the next clock cycle. Once the slave has indicated that it has completed the request, the bus will update the status of the original master request. If the master request was a burst request and it is not yet completed because there is still more data to transfer, the bus will update the address pointer within the burst request and retain the request for completion in subsequent cycles.

Normally the arbiter selects the request with highest priority on each cycle. However, if during the last cycle the request that was executed had its "lock" flag set when the master issued the request to the bus, then:

1. If the request in the previous cycle was a burst request and it is not yet completed, then it is always reselected.

2. If the master that issued the request in the last cycle is issuing another request in the current cycle, then that master's request is always selected.

As you can see, burst requests are normally interruptible by higher priority bus requests, but the bus locking feature makes it possible to specify that they not be interrupted. The bus locking feature also makes it possible for masters to implement atomic bus synchronization transactions for CPU instructions such as "test and set."

8.7 Two-Phase Synchronization

We call the overall synchronization scheme used within the `simple_bus` design a *two-phase synchronization scheme.* This is because certain modules (masters and slaves) are active on the rising edge of the clock, while other modules (bus and arbiter) are active on the falling edge.

The two-phase synchronization scheme was chosen because it is simple and it works well for transaction-level models of busses. It also would have been possible to make the bus be a primitive channel and to have used the `request_update/update` scheme described in chapter 7, but the two-phase scheme was chosen because:

- We can make the bus be a `sc_channel` instead of a `sc_prim_channel`, and this allows the bus to have a process and a port within it.

- Because the bus now executes on the falling edge of the clock, we can be sure that by the time the bus executes it has gathered all of the requests for this bus cycle, since all masters execute on the rising edge. It is even possible that some of the masters may be signal-level models that deliver their requests to the bus some number of delta-cycles after the rising edge. In the latter case, it won't matter that there are varying numbers of delta-cycle delays when the requests are presented to the bus, since the bus won't execute until the falling edge of the clock.

In chapter 7 we saw that primitive channels and the `request_update/update` scheme provided a clean method for constructing deterministic designs in SystemC. Now we can point out some strong similarities between the two-phase synchronization scheme and the request_update/update scheme and see that the two-phase synchronization scheme also provides a convenient approach for constructing deterministic designs:

- In the two-phase scheme, all communication between modules attached to the bus goes through the bus. In the request_update/update scheme we saw that all communication between processes ultimately go through primitive channels that use request_update/update.
- In both schemes all communication is delayed by some amount of time (a clock cycle in the two-phase scheme, a delta-cycle in the request_update/update scheme.)
- In the two-phase scheme, on the rising clock edge, all bus requests are delivered to the bus and stored. No state changes of the bus are externally visible at this point. This is similar to the SystemC evaluate phase used in the request_update/update scheme—in this phase state changes to primitive channels are not externally visible.
- In the two-phase scheme, on the falling clock edge, the bus deterministically arbitrates the requests, selects a single one for execution, and transfers data to/from the appropriate slave. This is similar to the update phase of the SystemC simulation cycle—in this phase primitive channels resolve or arbitrate competing requests.
- In the two-phase scheme, a master that initiates a read request to a slave in the current cycle will only see the data in the next clock cycle, and only if the request has completed. In the request_update/update scheme, we saw that one process receiving data from another process via a primitive channel would see that data at the soonest within the evaluate phase within the next delta-cycle.

It should be noted that the use of the falling clock edge for triggering the bus is just a modeling technique. It does not imply that an actual implementation of the design would also use the falling edge. In an actual implementation, the bus functionality might be merged with the blocks attached to the bus such that all sequential logic is positive edge triggered.

Designs that adhere to the two-phase scheme will be deterministic since the order of process execution within the phases will not affect execution results, as long as arbitration of bus requests during the falling edge phase is performed in a deterministic manner. (In `simple_bus`, arbitration is completely deterministic because all masters have unique priorities.)

Both the two-phase scheme and the `request_update/update` scheme have their respective strengths. The two-phase scheme is very efficient for bus-based communication and it provides modeling flexibility because the bus channel itself can have a process within it. The `request_update/update` scheme is very similar to what HDLs provide, works well for both synchronous and asynchronous logic, and is easy to use because designers generally don't need to consider the global design when using it. The fact that you can use both approaches in SystemC, and even create designs that use each scheme in different parts of the design demonstrates the modeling flexibility available within SystemC.

8.8 Common Questions

It seems that the distinction between modules and channels is being blurred within `simple_bus`. What's the reason for this?

It is true that the distinction between modules and hierarchical channels is blurred within `simple_bus`. In previous chapters we've seen how channels generally are used to model communication, generally do not have ports, and generally implement interface methods. In contrast, modules generally have ports to access channels and modules generally do not implement interface methods.

But in `simple_bus` these generalizations no longer apply. We see that the bus itself is a channel that implements channel methods, but that it also has ports like a module. And we see that the slaves and arbiter, rather than being modules, are channels that have no ports but which implement channel methods.

What's going on? First, from a language perspective, hierarchical channels (`sc_channel`) and modules (`sc_module`) are the same thing. SystemC declares `sc_channel` to be a `typedef` for `sc_module`. Thus, any distinction between modules and channels really only exists in the user's mind, and is not significant from a language perspective. Informally we say that channels implement one or more interfaces whereas modules do not, but even this distinction doesn't work well within `simple_bus`.

In the `simple_bus` design we have used the equivalence between modules and channels as a way to optimize the overall design with the goal of minimizing the total number of processes within the design and the number of context switches between processes as the design simulates. The overall strategy is to design blocks that implement (or "service") transactions as channels that inherit from, and implement, their transaction interfaces, while we make blocks that initiate (or "use") transactions to have normal ports that allow them to access the channels that provide the implementation of those transactions. Since the master blocks only initiate transactions, we can design them as normal SystemC modules. But all of the other blocks in the design end up being channels since they all implement one or more transaction interfaces. The bus channel within `simple_bus` is the extreme case—it implements several transaction interfaces, and it also has ports that it uses to access the transaction interfaces of the slaves and arbiter.

Why do slaves implement the `simple_bus_slave_if` rather than having normal ports like most other modules? And why does the arbiter implement the `simple_bus_arbiter_if` rather than having a normal port?

The answer is that this allows us to totally eliminate the need for a process within the `fast_mem` and `arbiter` modules, and it allows us to minimize the amount of work that the process within `slow_mem` does. Consider the `very_simple_bus` example presented above. If we relocated the `_mem` data array within `very_simple_bus` to a separate channel, and then had `very_simple_bus` read and write data to that data array via channel methods, we would have a model that mirrors what we have for `fast_mem` within `simple_bus`. Note that relocating the data array to a separate module in this case does not necessitate the introduction of another process within that separate module, since the data array is still accessed in a sequential manner without any simulated delay. Similarly, there is no need to introduce a process within the arbiter since a channel method within the arbiter (invoked by the bus process) can perform the arbitration without any simulated delay.

In addition, consider that the slave and arbiter modules only respond to requests rather than initiating them. By choosing to have these modules implement the methods of the `simple_bus_slave_if` and `simple_bus_arbiter_if`, we can then define the methods within these interface classes that directly correspond to the transactions that the slaves and arbiter implement.

Why are multiple slave channels attached to the same slave port on the bus?

We could use a separate sc_port object within the bus for each slave that is attached, but that would mean that the model for the bus would then have a fixed number of slaves, equal to the number of such ports.

Since we don't want to fix the number of slaves, we instead take advantage of the SystemC *multiport* feature. Usually exactly one channel is bound to each port when a design is elaborated in SystemC. However, by adding an additional integer argument to the sc_port template parameter list, we can indicate that more than one channel can be bound to a single port object. (If this additional argument is greater than one, then it specifies the maximum number of channels that can be bound to the port. If the argument is zero, then any number of channels can be bound to the port.) Within the simple_bus class definition above, we see that we define the slave_port as:

```
sc_port<simple_bus_slave_if, 0> slave_port;
```

This allows us to bind as many slaves to the bus slave_port as we wish during elaboration. During simulation, the bus can determine the number of channels bound to the slave_port using the call slave_port.size(), and the bus can access each of the separate slave channels bound to slave_port by using the syntax slave_port[N] where N is the index of a particular slave. The indexing order is guaranteed to match the order in which the interfaces were bound to the multiport during elaboration.

What if you want to create a master module that uses more than one of the three master interfaces introduced above? For example, what if a master module needs to use both the burst_write() function and the write() function?

In this case, combine the separate interfaces into a new unified interface class:

```
class simple_bus_unified_master_if
  : public simple_bus_direct_if,
    public simple_bus_non_blocking_if,
    public simple_bus_blocking_if
{
    // empty
};
```

Now make simple_bus inherit from simple_bus_unified_master_if rather than from each of the three separate interfaces:

```
class simple_bus
  : public simple_bus_unified_master_if,
    public sc_module
{
    // as before
};
```

Now masters can use either of the four interface classes according to their needs. If a master module uses the `simple_bus_unified_master_if` interface, it will have access to all of the methods within all of the interfaces.

8.9 Executing and Experimenting with the Design

You can compile, build, and execute the `simple_bus` design without making any modifications to it. When you execute it, you will see monitor output from the third master module, which uses the direct master interface to monitor the contents of some memory locations within the design. The first and second masters will generate read and write transactions that cause the state of the memory to change as the design executes.

A good way to understand the execution of the design is to turn on the verbose option within the bus itself. This can be done by setting the verbose flag to true when the bus constructor is called. You will see logging output that indicates what requests are being sent to the bus, and what requests are selected by the bus at various times. Another good way to monitor the execution of the design is to turn on the verbose option to the arbiter module. This will allow you to track which request it selects among the competing bus requests.

Here are a few example design experiments that could be done with the `simple_bus` design or other similar transaction-level models:

- You can add different numbers and kinds of masters and slaves.
- You can change the priority of masters and change the arbitration rules that are used.
- You can experiment with using or not using the bus locking feature.
- You can change the number of wait states for slave operations to complete.
- On masters that represent CPUs, you can execute a high level software model that represents the software intended for the actual sys-

tem. You can annotate the software code with `wait(sc_time)` statements to account for estimated execution delays of the software. Calls to device driver routines should be replaced by code that calls the interface methods of the bus channel. For such software models you can also add additional masters onto the bus to model bus traffic expected due to CPU instruction fetches and reading and writing of data. These models can use a statistical traffic generation scheme or other approaches to generate the expected bus traffic.

- You can link an instruction-set simulator (ISS) model into the SystemC simulation to model software running on a CPU or DSP. Generally such a model would use the nonblocking master interface. This would then allow you to execute actual object code (including RTOS and assembly code) for the software part of the system along with the transaction-level hardware models. In addition, you can also use the direct master interface to enable a software debugger to run on top of the ISS as discussed in section 8.4.

- You can create cycle-accurate or cycle-approximate models of hardware modules in SystemC and attach them to the bus as slaves to allow them to be modeled along with the rest of the system. Then you can refine these hardware modules completely to RTL within the same SystemC environment, reusing the original testbench. You can also attach pin-accurate hardware to the bus using *adapters* or *converters*. This will be explained in section 9.5.

8.10 Measuring Simulation Performance

As we noted at the beginning of the discussion of `simple_bus`, high simulation speed is a key goal. We've already shown a number of the modeling techniques that enable high simulation speed, but you may be wondering what its actual performance is. The answer is that the `simple_bus` design is capable of running at about one hundred thousand clock cycles per second on typical PCs and workstations. To see this level of performance, you will need to use both an optimized (i.e., nondebug) version of the SystemC library and also of the `simple_bus` design itself, and you will need to disable all logging output so that it doesn't affect the timing result.

It's fair to ask whether such high performance numbers will carry over to other larger and more complex designs. To a large extent this depends on the modeling style that is used. If the transaction-level modeling style and the high speed techniques used within simple_bus are applied rigorously throughout the system, roughly similar performance levels may still be reached despite the larger size of the design. This is possible because on most clock edges, only a few blocks actually execute, and even fewer do any significant work.

Of course, if models are added to the design which do not follow these high performance techniques, performance will be much worse. All it takes is a single slow model (e.g., an RTL model) added to the design to dramatically slow down the simulation speed.

8.11 Summary

We have introduced the transaction-level modeling style and shown how it allows designers to model communication in an abstract way, by separating the implementation of communication functionality from those modules that use the communication functions. We have seen how this separation makes models easier to use because models are simpler, and also enables significantly higher simulation performance when used throughout a design. These benefits make transaction-level models very useful in functional modeling (both timed and untimed), platform modeling, and testbenches.

With the simple_bus example, we have seen how to construct high-performance, cycle-accurate, transaction-level models of platforms in SystemC. Such models are:

- Relatively easy to develop, understand, use, and extend
- Able to accurately model both the hardware and software components of a system
- Capable of being constructed very early in the system design process, thus enabling designers to explore implementation alternatives and make design trade-offs before it's too late or too expensive to do so
- Fast and accurate enough to validate software before more detailed hardware models or implementations are available

Because of these important features, we expect the use of transaction-level models to increase in the future.

In the next chapter we will see how transaction-level models of platforms such as simple_bus can also be used within the communication refinement flow.

9

Communication Refinement

9.1 Introduction

In the previous chapters we have discussed how modules and channels at different levels of abstraction are described in SystemC. In the process of transitioning from an abstract model to a more detailed specification, we not only refine the model's internal structures, its timing, or the data-types being used—we also need to think about how this component exchanges information with its environment.

Communication refinement refers to expanding an abstract communication protocol into an actual implementation. During this process the abstract communication scheme is either mapped onto whatever is provided by a given target architecture, or, constraints of the target architecture permitting, refined into an efficient custom implementation.

An example of an abstract communication protocol is a high-level FIFO (first-in-first-out) scheme as implemented by the `sc_fifo` channel where modules communicate via blocking read and write operations. Due to the blocking nature—for instance, a data producer is suspended until there is space available in a buffer—control is implicit, which makes this communication scheme very easy to use in high-level functional models.

This abstract protocol can be refined into either hardware or software. A hardware implementation may consist of a shift register or a RAM together with some control logic, while a software implementation may use a certain memory range in conjunction with a modulo addressing scheme.

SystemC allows for the design and implementation of abstract communication protocols. The key to this is the use of interface method calls (IMC, cf. section 2.4). To model an abstract communication protocol,

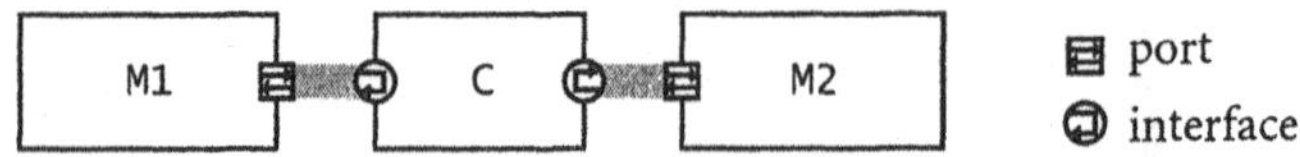

Figure 9.1: Modules M1 and M2 communicate via the channel C

users first describe the interfaces of a communication protocol. Then, primitive or hierarchical channels implementing these interfaces can be designed.

Communication *refinement* sounds harmless—but indeed it is a complicated design task; there is no simple one-size-fits-all approach to it, and hence this chapter cannot deliver one. Also, the focus of this book is the SystemC language—not on a particular methodology. In fact, SystemC supports a number of different methodologies. This chapter emphasizes the language-oriented aspects of communication refinement, exemplifying what can be done *today* with SystemC. Although a number of design tasks presented in the sequence can be automated, we make no assumptions regarding the use of any commercial design tools.

After presenting the basic steps within the communication refinement process, we will next take a look at the communication refinement process in a more general way. We will apply the communication refinement steps to different communication scenarios. We begin with refining a very simple example using abstract FIFO communication into a hardware implementation. Next, we map the same example onto a software implementation. Finally, we address the case of a mixed hardware–software implementation.

9.2 Steps in the Refinement Process

Before we can take a closer look at specific examples we need to address the process of communication refinement in a more general way in order to introduce some key terms and ideas.

Let us start with assuming we have two high-level modules M1 and M2 communicating over a channel C via some abstract protocol (figure 9.1). One possible approach (by no means the only one) to refining this setup towards an implementation is to carry out the following steps.

1. Select an appropriate communication scheme that will be used in the implementation.

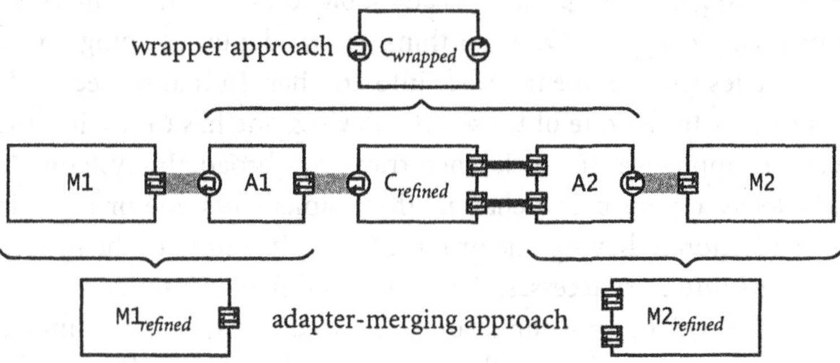

Figure 9.2: Alternative clustering scenarios; the footprint of $C_{refined}$ was chosen arbitrarily

2. Replace the abstract communication channel C with a refined one that implements the selected scheme ($C_{refined}$).

3. Enable the communication of the modules M1 and M2 over $C_{refined}$ by either

   (a) wrapping $C_{refined}$ in a way that the resulting channel $C_{wrapped}$ matches the interface needs of M1 and M2, or by

   (b) refining M1 and M2 into $M1_{refined}$ and $M2_{refined}$ such that their respective interfaces match $C_{refined}$ (adaptation).

As we can see, one outcome of step 2 is that the interfaces of the modules and the refined communication channel no longer match. The goal of step 3 is to undo this mismatch by modifying one of the connection points; either the channel is wrapped in a *wrapper module* to give it the appearance of the original channel, or the modules are refined so that they become compatible again (see figure 9.2). (We have already seen an example of the wrapper-based approach in section 7.5 with the `hw_fifo_wrapper` example.)

The two alternatives are quite similar in the sense that they both contain a refined channel and they both map the high-level protocol used in the original setup to what is provided by the refined channel. In the wrapper-based approach this mapping is achieved in the wrapper enclosing the refined channel. Alternatively, the mapping eventually takes place in the modules themselves.

In order to get to this point, it is advisable to carry out an intermediate step involving *adapters*. One can think of an adapter as being a module that translates (maps) one interface into another. In that respect its function is similar to the one of those little devices one has to use in order to get electric appliances to work when traveling abroad. In SystemC terms an adapter is a hierarchical channel that implements one or more interfaces in addition to having one or more ports. It translates the interface it implements into port accesses, that is, into a different interface. Referring back to the `hw_fifo_wrapper` example in section 7.5, we can think of an adapter as a separate channel that contains either the implementation of the `read()` or `write()` methods but nothing more.

The introduction of adapters as an intermediate step has a number of definite advantages. Adapters are modules that can be easily reused. In practice we find that typically only a finite number of different interfaces are being used at both the algorithmic and architectural level. Hence, a small library of adapters is usually sufficient to cope with the needs of even a larger design project.

Furthermore, the combination of adapter and high-level module is often "good enough" for the first design steps to be carried out after selecting a refined channel, such as verifying the functionality of the system or analyzing its performance by means of simulation.

As the design effort progresses there are basically two things that can be done with an adapter. One can introduce an additional level of hierarchy by wrapping the refined channel and the adapters connected to it. We wind up with $C_{refined}$ being wrapped in such a way that the resulting channel $C_{wrapped}$ has the same interface as the original channel `C` (step 3a, figure 9.2).

Alternatively, one can merge the adapters into the calling modules (i.e., the modules containing the calling processes) resulting in refined modules $M1_{refined}$ and $M2_{refined}$ (step 3b, figure 9.2). This step becomes particularly easy when only a single module accesses the adapter's interface. In this case one can simply gut out the adapter, move all members (ports, processes, methods, data fields) over to the calling module and replace calls to the adapter's methods with what has now become a new method of the calling module. We will see an example of this later in this chapter.

Whether wrapping or merging is the right thing to do depends to a large degree on the chosen target architecture and on the specific methodology. Later in this chapter we will take a closer look at different scenarios (hardware–hardware, software–software, and hardware–software) and we will detail which technique is preferable in the different cases.

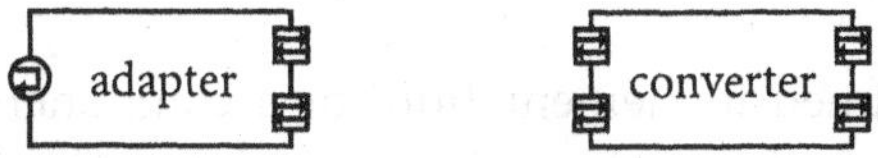

Figure 9.3: Different *footprints* of adapters and converters

But before we do that let us remain at the conceptual level and examine how a channel can change its *footprint*—the ports and interfaces of a module—in the process of refinement. Speaking in terms of SystemC, a channel is distinguished by the fact that it implements an interface; that is, the channel has a set of *interface methods* (cf. section 2.4) that can be accessed through the ports of connected modules. In the refinement process, the set of interface methods for a channel may change. Alternatively, a module that does not implement any interface methods at all but has ports instead might replace a channel. It should be noted that introducing adapters (optionally followed by a merging or wrapping step) could be applied in both cases.

But what if a refined model with a different set of ports replaces one of the abstract modules itself? This may happen if, for instance, the refined model is a piece of existing IP that is to be reused. In order to make the refined module and communication channel fit together a *converter* may be required.

The main difference between an adapter and a converter is that the adapter is a hierarchical SystemC channel while the converter is a "regular" module that has "ports on both sides" (see figure 9.3). Making this distinction is sensible since adapters are often only temporary objects that in the course of the refinement process get merged into the calling modules and, hence, disappear. (Testbenches are an exception in that it is not necessary to merge adapters there.) Converters, in contrast, tend to remain as a component in the final implementation.

Other than that a converter serves a purpose similar to an adapter. The result of wrapping a refined module and a converter is a new module that matches the channel's I/O footprint. In some cases we can see that a converter accesses an adapter through the adapter's interface. These can then be merged in order to form a new, refined converter. Of course, converters can be reused just the same way as adapters.

Having introduced the concepts of wrapper, adapter, and converter, we can now describe the general process of communication refinement as a repeated execution of the following steps:

1) Select & replace

Select a refined implementation for a component in your system. This can be either a channel or a module. The choice of whether to pick a channel or a module in this step is largely determined by the target architecture and the availability of existing IP. Replace the original component with the refined component.

2) Insert adapter or converter (optional)

In order to make modules and channels play together again, we need to insert either an adapter (channel refinement) or a converter (module refinement). Since both have a high potential for reuse they should be stored in a generally accessible module library. Of course, this step is not required if interfaces have not changed during the first step.

3) Analyze I/O functionality and performance

After adapters and/or converters have been inserted into the system, it is operational again, that is, the (possibly refined) modules of the system can communicate via the (possibly refined) channels once again. The resulting setup can often be used for simulating the system's performance, for instance, in terms of bus load or latency. Of course, if results are not satisfactory, either one or both of the last two steps may have to be carried out again.

Before we analyze the system's performance, we need to check its functionality. Although in terms of the interfaces and protocols the system is operational again, there is no guarantee that the behavior remains unchanged.

Imagine a case where modules originally communicated over abstract channels using blocking read and write operations. If, for instance, the refined channels have a smaller effective storage capacity at some point in time (for example, this can happen if the refined channels support only single read or write operations per clock cycle while the abstract counterpart could be accessed more than once within a single delta-cycle), then read or write operations that did not block before may do so now. This may result in deadlocking the complete system. And, of course, there is always the chance of a faulty implementation.

4) Flatten (optional)

A channel might have been replaced with a wrapped version of a refined channel (step 5 below). Flattening (ungrouping the hierarchy of the wrapped channel) makes the adapters visible, resulting in a potential for merging adapters and connected modules.

5) Wrap (optional)

We can wrap a channel and the set of connected adapters in order to create a new channel that encapsulates the refined channel. As we will see in later sections of this chapter, this approach is sometimes helpful when targeting software implementations.

6) Merge (optional)

If adapters are present in the system, they can be merged into the calling module. This step is sometimes referred to as *protocol inlining* [4], although the term is misleading as it often requires more than just inlining some access functions into the calling module's code. This step is definitely required in the process of refining an abstract communication protocol towards a hardware implementation (refinement of a transaction-level model to a pin-level model).

7) Restructure control flow (optional)

After merging adapters it may be necessary to restructure the control flow of the newly created module. This need arises mainly when refining high-level modules whose functionality was described in a sequential manner into a hardware implementation. Parallelism inherent to the underlying algorithms often needs to be made more explicit in order to achieve a good quality of result.

Consider an untimed functional model that uses blocking write operations. We may find code such as

```
...
first_output->write(some_data);
second_output->write(more_data);
...
```

Only after writing data to the first output is completed do we then begin to write data to the second output. The refinement steps described so far do not alter this behavior. This sequential execution scheme might not be suited for a hardware implementation, though. We may want to refine the control flow in order to exploit more parallelism.

```
...
first_port->initiate_transmission(some_data);
second_port->initiate_transmission(more_data);
wait_for_completion(first_port, second_port);
...
```

The pseudo code above shows that both transmissions can run in parallel if we are able to split a blocking I/O operation into an initiation phase and a second part where we wait for completion. A similar refinement procedure might be performed to enable channel accesses that are initially sequential to be performed in a pipelined manner.

We will give an example of control flow restructuring later when we address the refinement to hardware implementation.

8) Analyze Implementation (optional)

After flattening, wrapping, and merging (all of which are optional), the system might be in a state that allows an analysis of its implementation characteristics. For instance, parts of the subsystem may have been refined to a hardware implementation that can be fed into logic synthesis or behavioral synthesis tools so that cost factors such as area, latency, clock frequency, and power dissipation can be analyzed.

It may be required to check both functionality and I/O performance again during this step. Design errors may have been introduced during some steps. Also, restructuring the control flow (step 7) certainly makes it necessary to verify the new implementation.

Having reviewed the general steps in the communication refinement process, we can now apply these techniques in different scenarios. We begin with looking at the communication between two hardware modules.

9.3 Hardware–Hardware Communication Refinement

Figure 9.4 depicts the simple example system we will be using throughout the rest of this chapter. Two functional models, SOURCE and SINK, communicate via an abstract FIFO channel. The FIFO channel is an instance

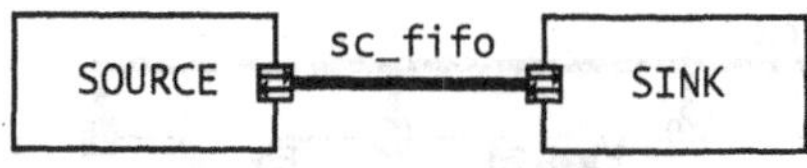

Figure 9.4: Untimed functional models communicating over a FIFO channel. The primitive channel will later be replaced with a refined version.

of the primitive SystemC channel sc_fifo. The SOURCE and SINK modules each contain a single thread process (SC_THREAD) that accesses the channel through the respective ports using an abstract communication protocol consisting of blocking read and write operations.

We begin the refinement process by selecting a hardware implementation of a FIFO channel that will replace the sc_fifo instance. Note that we could also have decided to start refining the SOURCE or SINK module.

Figure 9.5 shows that we selected a clocked hardware module HW_FIFO that implements a simple bidirectional handshake protocol where data is read (written) at the rising clock edge if both the valid and the ready lines are asserted. In order to connect SOURCE and SINK to HW_FIFO, we need a pair of adapters that transform the FIFO's input and output interfaces into pin-level accesses. The relevant parts of the FIFO_WRITE_HS adapter are shown below. Note that HW_FIFO was already presented in section 7.5 and that FIFO_WRITE_HS closely corresponds to the implementation of the write() method within hw_fifo_wrapper also presented in that section—in this case we have separated the implementation of the write() method into a distinct module.

Example 9.3.1 *Adapter converting FIFO interface into a handshake protocol*

```
// This adapter converts the FIFO write interface into a bi-
// directional handshake protocol. It is a hierarchical channel
// that implements the sc_fifo_out_if interface. FIFO_WRITE_HS
// has four ports; one for the data, one for the clock, and two
// for the control lines

template <class T>
class FIFO_WRITE_HS
  : public sc_module,
    public sc_fifo_out_if<T>
{
public:
    // ports
    sc_in_clk    clock;
    sc_out<T>    data;
```

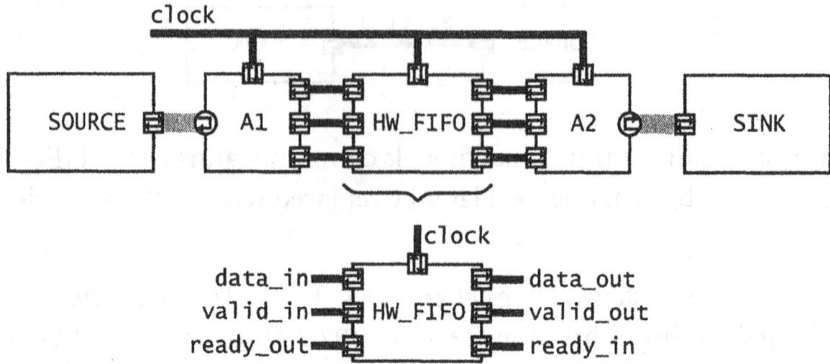

Figure 9.5: Example system after inserting HW_FIFO and appropriate adapters. A1 and A2 are instances of FIFO_WRITE_HS and FIFO_READ_HS, respectively.

```
    sc_out<bool> valid;
    sc_in<bool>  ready;

    // blocking write
    void write(const T& x) {
        data = x; // drive data line
        valid = true; // signal that data is valid
        do { // wait until data was read
            wait(clock->posedge_event());
        } while (ready.read() != true);
        valid = false; // make sure data doesn't get read twice
    }

    // provide dummy implementations for unneeded
    // sc_fifo_out_if<T> methods:
    bool nb_write(const T& x)
     { assert(0); return false; }
    int num_free() const
     { assert(0); return 0; }

    ...

};
```

When the write() method is called, the adapter drives the data line with the new sample and asserts the valid line. It then waits until the HW_FIFO model has read the sample. This happens on a rising clock edge if both the valid and the ready lines are active. A simple FIFO_READ_HS adapter can be implemented in a similar way.

Example 9.3.2 *SOURCE module before merging the adapter*

```
SC_MODULE(SOURCE) {
    // output port
    sc_fifo_out<int> output;

    // the one and only process
    void my_process() {
        int counter = 0;
        while(1) {
            counter++; // placeholder for decent functionality
            output->write(counter);
        }
    }

    SC_CTOR(SOURCE) {
        SC_THREAD(my_process); // declare process
    }
};
```

After inserting (but still before merging) the adapters as shown in figure 9.5, the system can be simulated again. In order to refine the SOURCE module to a hardware implementation, we can now merge the adapter into the SOURCE module (the "calling module," cf. step 6 above, page 159). The source code of the original SOURCE module is shown above before adapter merging, and the source code after merging the adapter is shown below.

In the process of merging the adapter into the calling module, we copy and paste the adapter's ports, methods (only the ones that are actually used), data fields (none in this case), and processes (none either) into a copy of the original module that we name HW_SOURCE. What used to be a method of the adapter now becomes a method of the refined module. We then replace the template argument (T) of the adapter with the data-type that is actually being used (int). Next, we remove the original port that was used to access the adapter, including all references to it. The resulting source code is shown below; changes are indicated in bold.

Example 9.3.3 *SOURCE module after merging the adapter*

```
SC_MODULE(HW_SOURCE) { // name changed
    // output port
    // sc_fifo_out<int> output // removed original port

    // new ports; imported from adapter
    sc_in_clk     clock;
    sc_out<int>   data; // replaced T with int
```

```
    sc_out<bool> valid;
    sc_in<bool>  ready;

    // blocking write
    void write(const int& x) { // replaced T with int
        data = x; // drive data line
        valid = true; // signal that data is valid
        do { // wait until data was read
            wait(clock->posedge_event());
        } while (ready.read() != true);
        valid = false; // make sure data doesn't get read twice
    }

    // the one and only process
    void my_process() {
        int counter = 0;
        while(1) {
            counter++; // placeholder for decent functionality
            // output->write(counter);
            write(counter);
            // replaced output->write() with write()
        }
    }

    SC_CTOR(HW_SOURCE) { // name changed
        SC_THREAD(my_process); // declare process
    }
};
```

Care must be taken that none of the "imported" ports, data members, methods, or processes causes a name clash with an existing name. If a name clash should occur, names should be made unique during the merge operation, for instance, by using the original port name as a prefix.

Furthermore, code executed as part of the adapter's constructor needs to be merged into the calling module's constructor(s). If the adapter module had additional constructor arguments, then these can usually be replaced by constant expressions. Clock ports of the adapter may require special treatment if the calling module was already connected to the same clock. Then, instead of creating a second clock port connected to the same signal, it is advisable to replace references to the adapter's clock with accesses to the calling module's clock.

The result is a refined pin-level module (HW_SOURCE) that can be used as a starting point for hardware synthesis. It can replace the SOURCE and FIFO_WRITE_HS instances in the system. (A final refinement step towards using static sensitivity where the process is made sensitive to the clock and

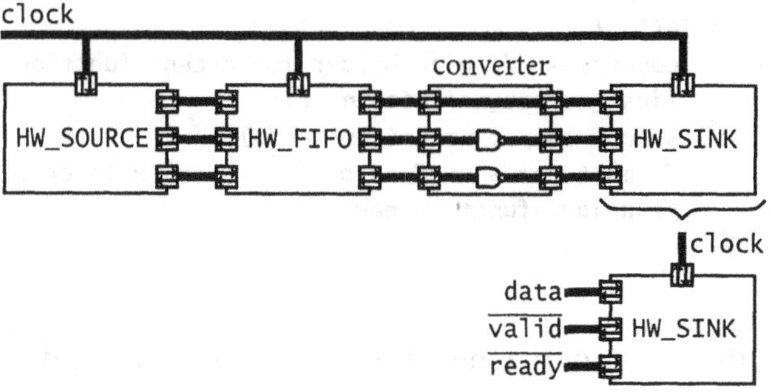

Figure 9.6: Example after refining all components; a converter is required as HW_SINK uses active-low inputs for the control lines

wait(clock->posedge_event()) is replaced with wait() may be required by a synthesis tool.)

We could pursue the same approach of refinement to a custom hardware module for the SINK module. An alternative scenario to consider is one where we select an existing hardware implementation (HW_SINK) and use a converter to connect it to the HW_FIFO module. Figure 9.6 depicts this scenario. The role of the converter is simply to reverse the polarity of the valid and ready lines.

Before we move on to software implementations, let's look at control flow restructuring (step 7 above). Assume that instead of having just a single output port, the SOURCE module instead had two outputs. Its one and only process might look like the one below.

```
// BEFORE REFINEMENT
void my_process() {
    int counter = 0;
    while(1) {
        counter++; // placeholder for decent functionality
        first_output->write(counter);
        second_output->write(counter/2);
    }
}
```

The refinement via adapters and merging would leave this process mostly in its original state.

```
// AFTER REFINEMENT
void my_process() {
```

```
        int counter = 0;
        while(1) {
            counter++; // placeholder for decent functionality
            first_output_write(counter);
            second_output_write(counter/2);
            // port names used as prefix in order to create
            // unique function names
        }
    }
```

Note that the names of the interface methods have changed. In order to create unique function names we used the original port names as a prefix. The same had to be applied to the ports.

If we inline the code of these two methods we get the following.

```
...
first_output_data = counter;
// using original port names as prefix to avoid name clashes
first_output_valid = true;
do { // wait until data was read
    wait(clock->posedge_event());
} while (first_output_ready.read() != true);
first_output_valid = false;

second_output_data = counter/2;
second_output_valid = true;
do { // wait until data was read
    wait(clock->posedge_event());
} while (second_output_ready.read() != true);
second_output_valid = false;
...
```

We see that the sequential behavior of the original untimed functional model is preserved; we begin to write data to the second port only after we succeed in writing to the first port. Restructuring the control flow as shown below will lead to a more parallel implementation with a smaller latency (in number of clock cycles). It should be noted that if the designer merely wishes to explore whether this more parallel control flow would be advantageous without having to refine the code as above, he can use *fork* and *join* constructs to first model this parallel implementation at a high level (see *examples/systemc/forkjoin* in the SystemC distribution). Then, if the results look encouraging, the refinement outlined below can be done.

```
...
// Resulting code after control flow restructuring.
// Initiate BOTH transmissions:
```

```
first_output_data = counter;
first_output_valid = true;
second_output_data = counter/2;
second_output_valid = true;

// wait until both transmissions are completed
do {
    wait(clock->posedge_event());
    if (first_output_ready.read() == true)
        first_output_valid = false;
    if (second_output_ready.read() == true)
        second_output_valid = false;
} while (! ( first_output_ready.read() &&
             second_output_valid.read() ) );
...
```

9.4 Software–Software Communication Refinement

It should be noted that the support for modeling software with SystemC, especially software running under control of a real-time operating system (RTOS), will be improved in future versions of SystemC. Chapter 11 will provide an outlook. In this section we will focus on the communication refinement of a software implementation based on the language constructs available today in SystemC.

Generally speaking, the refinement task is the same as before. We need to map the abstract communication scheme onto whatever is provided by the given target architecture or refine it into a custom implementation.

The primitive elements that can be used to implement a software-based communication scheme are memory plus communication and synchronization primitives offered by the operating system (OS). Memory (plus a set of access methods) is usually sufficient in the case of static (compile-time) scheduling of tasks—this might apply to a dataflow algorithm (see chapter 5). OS support is required when SystemC processes are mapped to concurrent threads or processes that run asynchronously with their own thread of execution—this is often the case with control tasks.

Mapping functionality described as a SystemC module to a software implementation usually means getting rid of all SystemC language elements, since (cross-)compilers for embedded architectures do not support SystemC module declarations, process declarations, channels, or data-types; they may not even support C++ at all but just C.

Hence, before we can address the question of communication refinement we must look at what it takes to actually get a SystemC module cross-compiled in the first place. This, of course, depends heavily on the target architecture in terms of compiler and operating system support. Below we will outline an approach that does not require more than a C compiler and an operating system with minimal thread support (without loss of generality we will be using the POSIX *pthread* interface [6]). It is targeted at refining a (timed or untimed) functional SystemC model towards a software implementation such that the *same* source code files can be used in a SystemC simulation as well as for cross-compilation with a plain C compiler. This has the clear advantage of not having to maintain the consistency of two completely separate design representations while allowing for the verification of the functionality in the system context at any point in time at almost no extra cost. (In reality, the coding overhead associated with the approach presented here is minor—although it looks more severe in the case of our simple example.)

We can think of a SystemC module as a number of processes, declared in SystemC syntax, embedded in a SystemC shell (the module) that potentially use SystemC data and channel types. Our approach will turn SystemC `SC_THREAD` processes into (POSIX) threads. (SystemC `SC_METHOD` processes can be turned into `SC_THREAD` processes by enclosing them in an infinite loop; within the loop body the main method of the `SC_METHOD` process is first called and then `wait()` is called.) SystemC data-types that are not available on the target architecture have to be replaced by built-in C data-types or `structs` as part of the refinement process. The module containing ports and data fields will become a C `struct` where ports are replaced by pointers. These, in turn, will be leveraged to map the SystemC interface method calls onto communication schemes available on the target architecture. If we apply these steps to the untimed functional implementation of the `SOURCE` module shown in example 9.3.2 we get the following header and source files.

Example 9.4.1 *SOURCE.h header file*

```
/* SOURCE.h */
#include "sc2sw.h"
/* replaces #include "systemc.h" */

#ifndef SYSTEMC
/* #include directives needed for cross-compilation */
#include "fifo.h"
#endif

struct SOURCE; /* forward declaration of SOURCE */
```

```
/* guts of processes */
void* SOURCE_my_process(struct SOURCE*);
/* this nonmember function becomes the actual process */

SC_MODULE(SOURCE) {

#ifdef SYSTEMC
    /* SYSTEMC => ports, processes, contructor go here */
    sc_fifo_out<int> output;
    void my_process() { SOURCE_my_process(this);}
    SC_CTOR(SOURCE) { SC_THREAD(my_process); }

#else
    /* C => equivalents for ports go here */
    struct fifo* output;
    /* this replaces the SystemC port */

#endif
    /* SHARED => data fields go here */
    /* int foo, bar;                 */
};
```

Example 9.4.2 *Contents of* `SOURCE.c`

```
/* SOURCE.c */
#include "SOURCE.h"

void* SOURCE_my_process(struct SOURCE* m)
{
    int counter = 0;
    while(1) {
        counter++;
        /* port access using an IMC macro */
        IMC1(m->output, fifo, write, counter);
    }
    return 0;
}
```

Example 9.4.3 *Contents of* `sc2sw.h` *header file*

```
/* preprocessor macros to ease cross-compilation */

/* include systemc.h only if SYSTEMC is defined */
#ifdef SYSTEMC
#include "systemc.h"
#else
#define SC_MODULE(X) struct X
#endif
```

```
/* Macros for calling interface methods with no, 1, or 2   */
/* arguments; can be extended as needed. The syntax can be */
/* polished if the preprocessor supports variadic macros.  */
#ifdef SYSTEMC

#define IMC(port,type,method) \
    port->method()
#define IMC1(port,type,method,arg1) \
    port->method(arg1)
#define IMC2(port,type,method,arg1,arg2) \
    port->method(arg1, arg2)

#else

#define CONCAT(a,b,c) a##b##c
#define IMC(port,type,method) \
    CONCAT(type,_,method(port))
#define IMC1(port,type,method,arg1) \
    CONCAT(type,_,method(port,arg1))
#define IMC2(port,type,method,arg1,arg2) \
    CONCAT(type,_,method(port,arg1,arg2))

#endif
...
```

The preprocessor macro SYSTEMC is used to distinguish between the use in a SystemC environment (in which case the code must be compiled with -DSYSTEMC) and C compiler environment (in which case -DSYSTEMC is not used).

Let us firstly look at the SystemC case. The SystemC header file is now pulled in via sc2sw.h. The main change is that the contents of member function my_process() have been moved into the nonmember function SOURCE_my_process(), which is called by my_process() in turn. The interface method calls have been rewritten using the IMC macros defined in sc2sw.h.

```
IMC(m->input, fifo, read)
```

comes out as

```
m->input->read()
```

if SYSTEMC is defined, resulting in the same interface method call as before. However, for cross-compilation it will be turned into

```
fifo_read(m->input)
```

where input is a pointer to a C struct. The corresponding declarations can be found in fifo.h.

Example 9.4.4 *Contents of the fifo.h header file*

```
/* C-based implementation of FIFO communication based      */
/* on communication primitives available under the target */
/* OS (e.g., based on pipes).                              */

struct fifo;

int fifo_create(struct fifo*);
int fifo_read(struct fifo*);
void fifo_write(struct fifo*, int);
...
void fifo_destroy(struct fifo*);
```

Basically, the SystemC interface method calls are translated into a set of C function calls that all share a common prefix (fifo_). The first argument is always a pointer to a corresponding data structure. This scheme allows one to easily establish a mapping between communication mechanisms supported by a target architecture (including operating system) and SystemC channels. Existing communication schemes found within SystemC can be encapsulated in a *C-based wrapper* such as the one shown in example 9.4.4 together with example 9.4.5, which may then be implemented using pipes or message queues provided by the RTOS.

Example 9.4.5 *Top-level code for the software implementation of the example shown in figure 9.4*

```
#define REENTRANT
#include <pthread.h>
#include "SOURCE.h"
#include "SINK.h"

int main(int argc, char** argv)
{
    struct fifo f;
    SOURCE source;
    SINK sink;
    pthread_t source_thread, sink_thread;

    /* initialize FIFO */
    fifo_create(&f);

    /* connect FIFO and modules /
```

```
    source.output = &f;
    sink.input = &f;

    /* create threads */
    pthread_create(&source_thread, NULL,
                   (void *(*)(void *)) &SOURCE_my_process,
                   (void*)&source);
    pthread_create(&sink_thread, NULL,
                   (void *(*)(void )) &SINK_my_process,
                   (void*)&sink);

    /* wait until all threads have terminated */
    pthread_join(source_thread, NULL);
    pthread_join(sink_thread, NULL);
    return 0;
}
```

In the *C-based wrapper* approach outlined above, we basically duplicate the functionality of SystemC threads and channels (sc_fifo in this particular case) within the C language running on top of an RTOS. An important alternative approach involves creating user-defined SystemC channels with an interface matching the equivalent C function calls of the target RTOS communication primitives. Both approaches lead to equally powerful yet slightly different refinement techniques. In the case where RTOS functionality is made available at the system level in the form of a SystemC channel, the refinement process is similar to the hardware–hardware case where in the end all communication schemes must be expressed in terms of a given set of SystemC channels (signals and resolved signals in the case of hardware, and channels that represent RTOS communication primitives in the case of software). Again, the concept of adapters and merge operations can be applied. The use of C-based wrappers (the first approach presented above) favors deferring the refinement until the implementation phase. While this draws a clear line between system simulation and implementation, it presents the danger that critical effects—such as inadequate buffer sizes stalling a complete system—may not be observed early in the design flow.

It should be noted that, although communication refinement is an important step in the process of turning a functional model into a software implementation, it is not the only one. Other important design tasks, which are clearly beyond the scope of this chapter, include data-type refinement (for instance, going from a description that uses floating-point arithmetic to a fixed-point representation suited for cross-compilation) and static scheduling. The purpose of static scheduling is to reduce the runtime overhead (e.g., for context switching) as far as possible by finding a (pseudo-)static execution order for a set of SystemC processes and merging them into a single thread. This technique is particularly well understood for dataflow modeling [5] [17] [18].

9.5 Hardware–Software Communication Refinement

For the sake of simplicity we have so far concentrated on the refinement of individual communication links in the previous sections. Now it makes sense to take a step back and take a look at the bigger picture. Purely untimed functional models typically use dedicated channels for intermodule communication. Shared communication links come into play when the mapping onto a target architecture is considered. Only then does the system model begin to reflect the communication infrastructure of the platform.

The first models used to analyze effects of shared I/O resources may still be abstract, such as the one shown in figure 9.7. In this example models communicate via point-to-point FIFO channels (sc_fifo) at the untimed functional level. A model of a shared communication link (XBAR, source code shown in example 9.5.1) has been introduced in the next step. Note that the individual models still use FIFO communication. The size of the FIFOs connected to XBAR becomes an important architectural parameter as it models the size of the memory of the input and output buffers interfacing with the crossbar.

The code presented in example 9.5.1 models XBAR using a simple clocked round-robin arbitration scheme. More complicated scenarios involving priorities, resource locking, wait states and so on can easily be modeled at this high level of abstraction. To allow an arbitrary number of entities to be connected, the XBAR module makes use of SystemC's ability to connect multiple channels to a single port. (This capability is enabled via the extra 0 argument to sc_port<>.)

Example 9.5.1 *Source code of an abstract crossbar model*

```
template <class T> SC_MODULE(XBAR) {
    // ports
    sc_port<sc_fifo_in_if<T>, 0> inputs; // arbitrary # channels
    sc_port<sc_fifo_out_if<T>, 0> outputs; // ditto: outputs
    sc_in<bool> clk; // the clock port

    // data
    unsigned last_; // index of last channel granted access
```

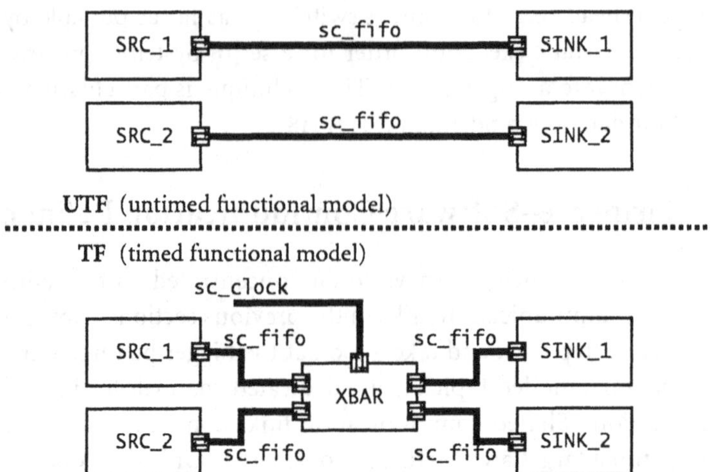

Figure 9.7: Modeling shared communication resources at a high level of abstraction

```
    // constructor
    SC_CTOR(XBAR) : last_(-1) {
        SC_METHOD(process);
        sensitive << clk.pos();
    }

    // methods
    void process() {
        // same # channels required
        assert(inputs.size() == outputs.size());
        // Iterate over all ports starting with the one that
        // comes after the one that was granted access last time.
        // Transmit data if both data available at the input and
        // space available at the corresponding output.
        for (int i=1; i<=inputs.size(); i++) {
            int current = (last_ + i) % inputs.size();
            if ((inputs[current]->num_available() != 0) &&
                (outputs[current]->num_free() != 0)) {
                outputs[current]->write(inputs[current]->read());
                last_ = current; // store info and we're done
                break;
                }
        }
    }
};
```

Introducing the target platform's communication infrastructure often goes hand in hand with annotating functional models with latency information (`wait(sc_time)`). Quite a number of architectural decisions can be analyzed at this level. The clear advantages of this approach include the ability to reuse untimed functional models as well as its ease of use and high simulation speed. Additional communication latency can be modeled by introducing a new FIFO channel that implements the same interfaces as `sc_fifo` but has nonzero delays.

After the system has been analyzed at a higher level of abstraction the platform is refined towards its final structure. This can be the outcome of a continuous refinement process or the result of selecting from a number of available IP blocks such as busses, memory subsystems, arbiters, bridges and so on. It is advisable not to immediately make a giant step towards a pin-level model of the platform but to firstly model it—possibly in a cycle-accurate way—at the transaction level. There are numerous advantages to transaction-level modeling (TLM, see chapter 8) at this point. The models are easier to use, develop, and maintain, compared to developing them at the pin level. Also, it is just a small step to refine (timed or untimed) functional models towards playing in concert with a TLM platform setup—essentially, because they are already transaction-based. This means that an executable model of the target platform can be made available early in the design flow, allowing for concurrent hardware and software design and verification.

Figure 9.8 depicts a refined version of the system depicted in figure 9.7 now using the `simple_bus` transaction-level bus model described in chapter 8. The `SRC` modules have become bus masters while the `SINK` instances are connected to the bus as slaves. (This choice has been made arbitrarily. The direction of the data flow does not impose a master-slave relationship. Hence, the `SINK` instances could have been implemented as bus masters, for instance.)

Relatively simple adapters like the one shown in example 9.5.2 are used to convert between FIFO and bus (or bus slave) interfaces. The potential for reuse of a parameterizable adapter—especially if it transforms a heavily used interface such as `sc_fifo_*_if`—is extremely high. Hence, such adapters should be stored in a easily accessed module library.

Example 9.5.2 *Adapter translating FIFO input operations into bus accesses*

```
// This adapter transforms a FIFO write interface into a bus
// access (bus master). It expects its counterpart (bus slave),
// which is registered on the bus under slave_address, to do the
```

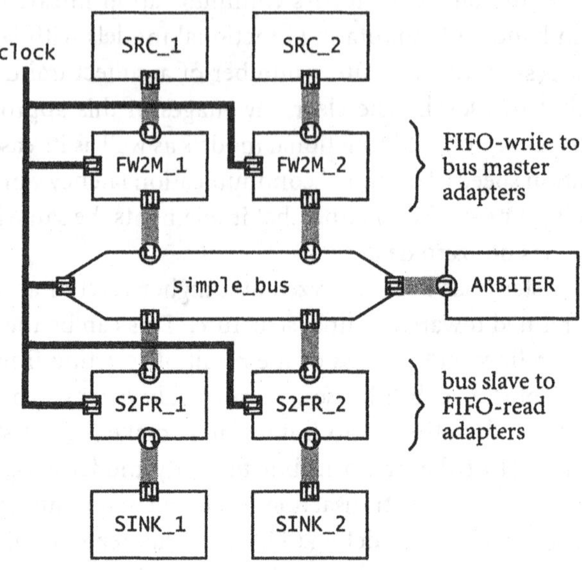

Figure 9.8: Transaction-level platform model

```
// actual buffering. If no space is available the slave will
// issue wait states. We can query available space by reading
// from slave_address. In order to avoid stalling bus traffic we
// wait for the specified number of clock cycles if no space is
// available and try again. Optionally, the bus is locked.
template <unsigned int slave_address,
          unsigned int unique_priority,
          unsigned int n_wait = 0, bool lock = false>
class FW2M : public sc_module, public sc_fifo_out_if<int>
{
public:
    // ports
    sc_in_clk clock;
    sc_port<simple_bus_blocking_if> bus_port;

    // blocking write
    void write(const int& data) {
        simple_bus_status status;
        int space_available;
        // check whether space is available
        for(;;) {
            bus_port->burst_read(unique_priority,
                                 &space_available,
                                 slave_address, 1, lock);
```

```
            if (space_available)
                break; // continue if there's space
            for (int i=0; i<n_wait, i++)
                wait(clock->posedge_event()); // else wait
        }
        // send data
        bus_port->burst_write(unique_priority, &data,
                              slave_address, 1, lock);
    }

    ...
};
```

It should be noted that there are a number of design alternatives to be explored at this stage; the following list includes just a few and is by no means complete:

- Connect a module to the bus as a master or as a slave.
- Send data immediately or use local memory to cache data and send it as a burst through the bus.
- If the bus is accessed as a bus master, then data can be sent to a memory (to be picked up by another bus master) or it can be sent directly to a slave (e.g., using memory-mapped registers).
- Additional side-band signals (e.g., interrupts) can be used for signaling purposes (e.g., to flag the availability or need for data). Alternatively, polling can be used for check for the presence of data.
- Depending on the feature set supported by the bus, it may be possible to lock the bus (if accessed as a master) or to initiate a split transaction (slave).

The design decisions to be made can have a significant impact on the overall system performance in terms of throughput, latency, and bus load. This makes it important to be able to execute and analyze a system model as early as possible in the design cycle—preferably using transaction-level modeling as described in chapter 8.

To verify correct operation of the hardware implementation we certainly need to model down to the pin-accurate level. Two approaches are possible (see figure 9.9). In both cases we have to replace the adapter connected to the respective module instance with one that transforms the FIFO access operations into a pin-level interface. Then, we could connect the module

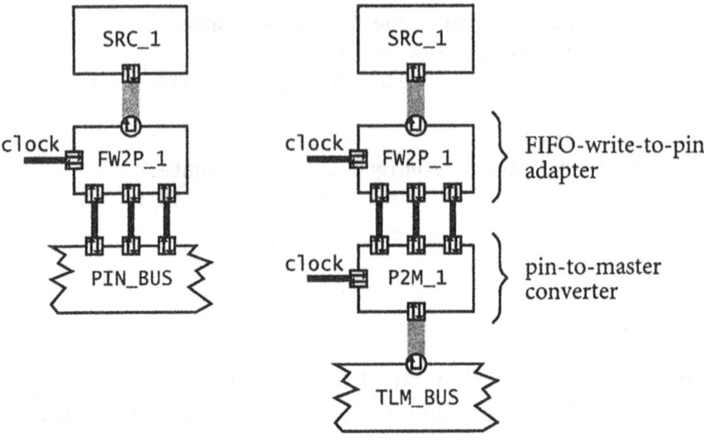

Figure 9.9: Two approaches to refining to the pin-level

and the adapter to a pin-level bus model. Alternatively, we could stay with the transaction-level bus model and pull in an additional converter that provides us with the required pin-level interface. This approach has two advantages. First, it allows for verifying subsystems in the system context before pin-level models of all components are available. Second, most of the blocks still operate at the transaction level resulting in a higher simulation speed.

In this section we have been focusing on hardware–software communication refinement, and up until now we have been discussing the hardware half of the refinement. Now let us focus on the software half. In the software–software communication refinement example presented in the previous section, it was not necessary to refine communication down to the bus transaction level since the existence of the target RTOS frees us from worrying about such details. But in hardware–software communication refinement, we must refine the software side down to the bus transaction level to be sure that it will interact properly with the hardware side.

Assume now that we wished to refine one of the SRC modules within figure 9.8 to a software implementation and therefore we needed to refine its interaction with the bus down to the bus transaction level. We would start by following the steps outlined in the software–software refinement flow above (section 9.4). We would not merge the adapter shown in example 9.5.2, though, as this particular one is more suited for a hardware implementation (e.g., it uses a clock signal). We could either search for a more appropriate adapter or address the software implementation directly at the untimed functional level. As explained in section 9.4, we would then replace the FIFO port accesses with IMC macros. This time, we could write

```
...
/* now using fifo_slave instead of fifo */
IMC1(m->output, fifo_slave, writer, counter);
...
```

The IMC macros allow for choosing arbitrary prefixes for the C structs and functions. We can now implement the necessary device-driver code using the fifo_slave prefix. A possible implementation is outlined in example 9.5.3.

Example 9.5.3 *Device driver for FIFO communication via the bus*

```
/* Just a rough sketch. Assumes RTOS support for nanosleep().  */

/* Data describing a connection to a FIFO slave. n_wait gives  */
/* the number of nanoseconds to wait in case the slave doesn't */
/* have free space.                                            */
struct fifo_slave { int* address, unsigned n_wait; };

void fifo_slave_write{struct fifo_slave* f, int data} {
    int space;
    struct timespec t;
    t.tv_sec = 0; t.tv_nsec = f->n_wait;
    for (;;) {
        space = *(f->address); /* read via bus */
        if (space) break;
         nanosleep(&t, NULL);
         /* a decent version would check the return value */
    }
    *(f->address) = data;
}
...
```

While the term *communication refinement* seemed sensible for the simple point-to-point examples discussed in the sections about hardware–hardware and software–software refinement, it may be more appropriate to speak of *communication design* in the case of hardware–software interactions. There is no single golden recipe to guide the designer through the numerous decisions that have to be made impacting an array of system parameters. What is *good* and *bad* can only be answered in the application context. Certainly trade-offs have to be explored.

9.6 Summary

In this chapter we firstly reviewed the different steps of a generalized communication refinement process. Then, we applied these techniques to simple examples of hardware–hardware, software–software, and hardware–software communication. We did not favor any particular methodology in our examples. Neither did we assume the availability of software tools other than a text editor and a C++ compiler—although some of the more mechanical refinement tasks lend themselves well towards automation. It is not feasible to automate the complete process. Communication refinement is often a challenging design task demanding a sound understanding of the architectures, algorithms, and application-oriented requirements. Anticipated product lifecycles and risk management are further aspects to be considered. This makes it hard to come up with one-size-fits-all approaches. However, two key principles can certainly be generalized: the use of well-defined and well-understood interfaces as well as the intelligent use and reuse of adapters and converters.

10

Testbenches, Tracing, and Debugging

10.1 Introduction

The main purpose of this chapter is to provide guidance on how to verify and debug a (possibly refined) design described in SystemC. This chapter is structured roughly according to the typical steps that have to be carried out in this process. First, we describe how to set up a testbench with a focus on refined, implementation-level models. Next, we give some hints on how the testbench can be instrumented in order to check a number of design properties. We then present ways to trace various sources of information and debug the system at a higher level. Finally, we provide some hints on how to debug a SystemC description at the source-code level.

10.2 Testbenches

The effort spent on setting up testbenches and verifying implementation-level models contributes significantly to the overall design effort. In some cases it turns out to be the single most resource-consuming activity in the complete design process. In order to keep cost as low as possible while achieving good test coverage, it is important to test intelligently.

As in other design areas, reuse and abstraction are powerful tools that should be leveraged whenever possible. In the given context, this means the reuse of abstract functional models when it comes to setting up test environments for lower-level models. The alternative of rewriting the complete system setup (including stimuli generation and data postprocessing) at the implementation level has a number of severe drawbacks. First, it is time consuming. Also, we would develop a second representation of a number of entities—a process in which design errors can be introduced. Hence, strictly speaking one has to verify that both representations are equivalent. Maintaining the consistency of these two sets of models can be very hard—especially if different design teams are involved. Finally, simulation speed is crucial in order to achieve the best possible test coverage making it all the more important to leverage abstract models whenever possible.

In testbench scenarios we find high-level models being used for different purposes: as stimuli generators, data postprocessors, reference models, or watchdogs (property checkers, monitors, observers). Figure 10.1 shows how a high-level functional model can be reused in creating a smart testbench. There, the functional reference model `REF` has been refined into what has now become the entity-under-test (`EUT`). The same stimuli generation (`SRC`) and postprocessing (`SINK`) models are still being used. In order to connect them to `EUT`, two adapters or converters (cf. chapter 9) are used. `H2L` converts the high-level interface into the implementation-level interface used by `EUT`. Similarly, `L2H` transforms the low-level interface into the high-level communication scheme used by the functional model. The functional reference model `REF` is still part of the picture: it is fueled by the same inputs as `EUT`. `DIFF` compares the two sets of output stimuli on the fly.

Checking behavior on the fly as part of the simulation run—as opposed to generating waveforms and comparing them afterwards with *golden output files*—is one technique for making testbenches smarter. Subtle design changes may result in changed waveforms requiring a time-consuming and error-prone "manual" inspection. Having smart tests that filter out harmless differences increases productivity and helps to create a stable set of regression tests. Also, as file I/O operations are time consuming it is often more efficient to check data on the fly compared to saving it in a file.

The idea behind property checkers, such as the `WATCHDOG` module shown in figure 10.1, goes along the same lines. For instance, it is not practically feasible to check whether a bus master correctly implements a bus access protocol at all times based on a manual inspection of register-transfer-level waveforms. An observer module can perform such tests with much less effort. Depending on the nature of the property to be checked, this module may be connected to abstract or refined interfaces . In figure 10.1 we connect `WATCHDOG` to both abstract and refined interfaces using dotted lines to indicate this.

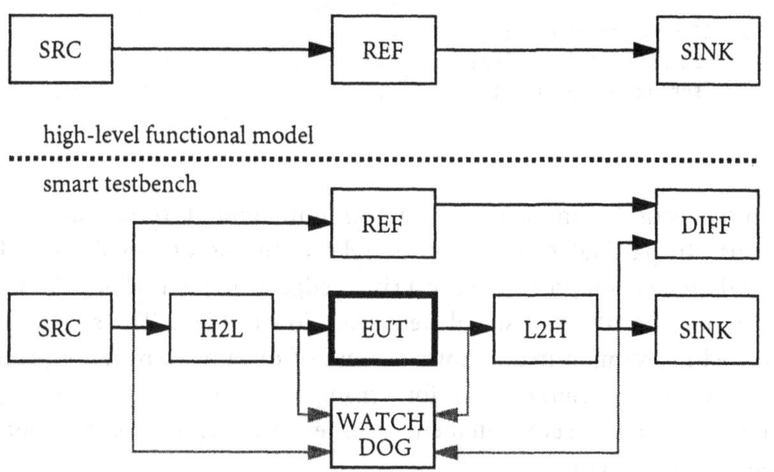

Figure 10.1: Different uses of high-level models in testbenches

SystemC offers a broad range of language elements that are useful for implementing observer modules. Static and dynamic process sensitivity often turns out to be a helpful tool. Let us look at two simple examples. The first example shows a module that checks the implementation of a handshake protocol. The observer (`hs_checker`) reports an error if the `data` line has glitches while the `data_valid` signal is asserted. It uses a method process (`SC_METHOD`) that is statically sensitive to the data line (see example 10.2.1).

Example 10.2.1 *Detecting glitches*

```
// We make this a class template such that the data signal
// can have any type.
template <class T> SC_MODULE(hs_checker) {
    sc_in<T> data;
    sc_in<bool> data_valid;

    void check_it() {
        // This process is triggered only if the data changes.
        // It complains about glitches if data changes
        // while data_valid is asserted. We have to excuse
        // the case where data_valid just changed itself.
        if (data_valid.read() && !data_valid.event())
            cerr << name() << ": glitch at t="
                 << sc_simulation_time() << endl;
    }
```

```
    SC_CTOR(hs_checker) {
        SC_METHOD(check_it);
        sensitive << data;
    }
};
```

In the second example we look at minimum and maximum delay conditions. In particular, we check the relative timing of two signals. The second signal is supposed to have a rising edge in an interval of [min_time, max_time] after the first signal went from low to high. The source code depicted in example 10.2.2 shows the use of dynamic sensitivity (calling wait() on an sc_event) to wait for a rising edge of first. It also leverages a more complex wait() statement in order to wait for a combination of events with a timeout.

Example 10.2.2 *Checking min and max delay conditions*

```
SC_MODULE(min_max_checker) {
    sc_in<bool> first, second; // first and second input

    // we need this because we do not use SC_CTOR
    SC_HAS_PROCESS(min_max_checker);

    sc_time min_time_, max_time_;  // min and max delays

    // constructor with name, min, and max delays
    min_max_checker(sc_module_name name,
                    const sc_time& min_time,
                    const sc_time& max_time)
        : sc_module(name),
          min_time_(min_time),
          max_time_(max_time)
    {
         SC_THREAD(check_it);
    }

    // helper function; prints error messages
    void print_error(const char* message)
    {
        cerr << name() << ": protocol violation at t="
             << sc_simulation_time() << " ("
             << message << ")" << endl;
    }

    // the one and only process
    void check_it()
    {
```

```
        while(true) {
            // wait for rising edge of first signal
            wait(first->posedge_event());

            if (second.read()) {
                print_error("second signal is already high");
                continue;
            }

            sc_time before = sc_time_stamp();

            // wait for either signal to change,
            // wait at most max_time_
            wait(max_time_, first->value_changed_event() |
                 second->value_changed_event());

            if (first.event()) {
                print_error("first signal went low too early");
                continue;
            }

            if (!second.event())
                print_error("delay overrun");

            sc_time delta = sc_time_stamp() - before;

            if (delta < min_time_)
                print_error("delay underrun");
        }
    }
};
```

It should be understood that the focus of examples 10.2.1 and 10.2.2 is on the use of SystemC language constructs for implementing certain types of runtime checks rather than on the particular checker modules themselves, which are on the lower end of the range of abstraction levels served by SystemC (cf. chapter 3). For instance, with only little modification the code shown in example 10.2.1 can be used to check that bus ownership is not taken away from a particular bus master during the course of burst transfer. Similarly, modules like `min_max_checker` (example 10.2.2) can check response times at all levels of abstraction.

Both adapters and converters (cf. chapter 9) can be used for protocol transformation (see `H2L` and `L2H` shown in figure 10.1). While adapters do have the advantage that they can be merged into the calling module in the process of communication refinement (protocol inlining), they tend to be slightly more complicated to develop (compared to converters) for new

SystemC users. For testbench purposes both adapters and converters work equally well.

In chapter 9 we have already seen examples of how to convert a high-level interface into a handshake protocol. In hardware testbenches we sometimes have to face the challenge that there is no true (bidirectional) protocol implemented at the register-transfer level. Instead, a subsystem works if and only if inputs or outputs are accessed at a predefined point in time. This can typically be found in computationally intensive data-path architectures where all explicit interface protocols have been optimized away and, instead, are implicitly designed into the timing of the subsystems. Often the timing pattern, that is, the points in time at which to serve inputs and outputs, is periodic after an initial setup phase. Example 10.2.3 shows a converter translating a value sequence read from a FIFO channel (`sc_fifo`) into hardware signal accesses using such a fixed timing pattern. The converter has been implemented as a class template in order to support arbitrary data-types. Its constructor takes four arguments: the instance name, the timing pattern, the interval length, and an initial offset. The timing pattern is an array of integers specifying when I/O is to be performed, that is, when data is read from a FIFO input and written to the hardware signal. The values specify the delay relative to the first clock edge. They must grow strongly monotonically. The pattern must be terminated by a negative value. The third constructor argument (`interval`) specifies the number of clock cycles after which the (finite-length) timing pattern is to be repeated. The interval must be larger than the largest value of the timing pattern. The offset argument specifies the length of an initial waiting period (before I/O starts). Nonzero values are required in cases where subsystems take multiple clock cycles for initialization after a reset.

```
int pattern[] = { 0, 2, 4, -1 };
f2s_pattern<int> converter("converter", pattern, 10, 20);
```

This code fragment shows the instantiation of a converter that transports integer numbers (see example 10.2.3, below). It uses a timing pattern of length 3 (the -1 just terminates the pattern, it is not part of it), an interval of 10 clock cycles and an offset of 20 clock cycles. Assuming a one nanosecond clock the converter's output will be written at

```
t = 20ns, 22ns, 24ns, 30ns, 32ns, 34ns, 40ns, 42ns, 44ns, ...
```

if data is available (the converter will issue an error message if not).

Example 10.2.3 *Converter translating a value sequence into hardware signal accesses with a fixed timing pattern*

```
template <class T> SC_MODULE(f2s_pattern)
{
    // ports
    sc_fifo_in<T> input;
    sc_in_clk clk;
    sc_out<T> output;

    // data fields
    const int* pattern_;
    unsigned int interval_;
    unsigned int offset_;

    // we need this as we do not use SC_CTOR
    SC_HAS_PROCESS(f2s_pattern);

    // contructor w/ name, timing pattern, interval, and offset:
    f2s_pattern(sc_module_name nm, const int* pattern,
                unsigned int interval, unsigned int offset = 0)
        : sc_module(nm), pattern_(pattern),
          interval_(interval), offset_(offset) {
        // check arguments; return without creating a process
        // in case of improper constructor arguments
        if (pattern == NULL) {
            cerr << name() << ": (ERROR) no timing pattern\n";
            return;
        }
        int max = -1;
        do { max = (max < *pattern) ? *pattern++ : -1; }
        while(max>=0 && *pattern>=0);
        if (max==-1 || max>interval_-1) {
            cerr << name()
                 << ": (ERROR) invalid timing pattern\n";
            return;
        }

        // create process
        SC_THREAD(main_action);
        sensitive << clk.pos();
        dont_initialize(); // we don't want the process to
                           // be triggered during initialization
    }

    void main_action() {
        for (int i=0; i<offset_; i++) wait(); // initial offset

        // periodic I/O as specified by timing pattern
```

```
        int counter=0, index=0;
        while (true) {
            if (counter==pattern_[index]) {
                do_io();
                if (pattern_[++index] < 0) index = 0;
            }
            counter = ++counter % interval_;
            wait();
        }
    }

    // helper function: do the I/O thing
    void do_io() {
        if (input->num_available() == 0)
            cerr << name() << ": (ERROR) input FIFO empty at t="
                 << sc_simulation_time() << endl;
        else
            output = input.read();
    }
};
```

10.3 Tracing

While tracing waveforms may not be the best-suited approach to functional verification and regression testing, it is certainly very helpful for debugging purposes. SystemC provides a set of functions that let us trace waveforms in different formats such as VCD (Value Change Dump), ASCII WIF (Waveform Intermediate Format), and ISDB (Integrated Signal Data Base). It is based on an extensible scheme; support for more data formats can be added.

Tracing a signal or a data member of a module consists of three steps

1. Create a trace file.
2. Register signals and variables to be traced.
3. Close the trace file before returning from `sc_main()`.

Example 10.3.1 shows this for the case of a VCD trace file. Replacing "`_vcd_`" with "`_wif_`" or "`_isdb_`" in the names of the functions to create and close the trace file will generate WIF and ISDB traces, respectively. Different types of trace files can be used together in one simulation run. Multiple trace files can be open at the same time. A signal or variable can be stored in any number of trace files simultaneously.

Example 10.3.1 *Writing signals and variables into a VCD trace file*

```
#include "systemc.h"

SC_MODULE(some_module) {
    sc_signal<int> sig; // signal and variable
    int var;            // to be traced

    SC_CTOR(some_module) {
        sig = var = 0;
        SC_METHOD(proc);
    }

    void proc() {
        sig = ++var;
        next_trigger(sc_time(10, SC_NS));
    }
};

int sc_main(int argc, char* argv[])
{
    // STEP 1: create trace file(s)
    sc_trace_file* tf = sc_create_vcd_trace_file("trace");
    // toplevel module
    some_module XYZ("XYZ");
    // STEP 2: register signals and variables to be traced
    sc_trace(tf, XYZ.sig, "XYZ.sig");
    sc_trace(tf, XYZ.var, "XYZ.var");
    // run the simulation
    sc_start(sc_time(500, SC_NS));
    // STEP 3: close the trace file(s)
    sc_close_vcd_trace_file(tf);
    return 0;
}
```

Trace files are typically created at the top level (inside `sc_main()`). However, it is possible also to create them in the constructor of a module or even create them while the simulation is already running. Signals and variables to be traced must be registered *after* the trace file has been created and *before* the simulation time is advanced. The functions creating trace files return a pointer to an `sc_trace_file` object, which is subsequently used as a reference when tracing objects or closing trace files (see example 10.3.1 and example 10.3.2).

Example 10.3.2 *Creating and closing trace files*

```
// create files trace.vcd, trave.isdb, and trace.awif
sc_trace_file* vcd_tf  = sc_create_vcd_trace_file("trace");
```

Ports	sc_in<T> sc_in_clk sc_in_resolved sc_in_rv<W>	sc_out<T> sc_out_resolved sc_out_rv<W>	sc_inout<T> sc_inout_resolved sc_inout_rv<W>
Signals	sc_signal<T> sc_signal_resolved sc_signal_rv<W>		

Table 10.1: Types of ports and signals for which SystemC supports tracing

```
sc_trace_file* isdb_tf = sc_create_isdb_trace_file("trace");
sc_trace_file* wif_tf  = sc_create_wif_trace_file("trace");
...
sc_close_vcd_trace_file(vcd_tf);
sc_close_isdb_trace_file(isdb_tf);
sc_close_wif_trace_file(wif_tf);
```

Only variables that have a lifetime that spans the complete simulation can be traced. This is true for all signals and data members of modules but local variables declared in functions cannot be traced.

SystemC comes with predefined support for tracing variables, ports, and signals of built-in C types and SystemC data-types. Table 10.1 shows the supported types of ports and signals. For built-in C types a bit-width *w* can optionally be specified—in that case, only the *w* least significant bits will be traced. (The behavior is undefined if a larger number of bits than are actually in the built-in type is specified—a VCD trace will show 'X' in that case.)

Example 10.3.3 *Tracing signals, ports, and data members*

```
SC_MODULE(my_module)
{
    // stuff to trace: ports, signals, data fields
    sc_in<int> input;
    sc_out<int> output;
    sc_inout<int> inout;
    sc_in_clk clk; // same as sc_in<bool>
    sc_signal<int> local_sig;
    int state;
    // constructor and other methods
    ...
};

int sc_main(int argc, char* argv[]) {
    sc_trace_file* tf = sc_create_vcd_trace_file("trace");
```

```
    my_module abc("abc");
    ...
    sc_trace(tf, abc.input,     "abc.input");
    sc_trace(tf, abc.output,    "abc.output");
    sc_trace(tf, abc.inout,     "abc.inout");
    sc_trace(tf, abc.clk,       "abc.clk");
    sc_trace(tf, abc.local_sig, "abc.local_sig");
    sc_trace(tf, abc.state,     "abc.state");
    // For signed/unsigned char/short/int/long we can also
    // specify a bit-width. Here, we trace everything again
    // as 2-bit.
    sc_trace(tf, abc.input,     "abc.input2",     2);
    sc_trace(tf, abc.output,    "abc.output2",    2);
    sc_trace(tf, abc.inout,     "abc.inout2",     2);
    sc_trace(tf, abc.local_sig, "abc.local_sig2", 2);
    sc_trace(tf, abc.state,     "abc.state2",     2);
    ...
}
```

Support for tracing additional data-types (user-defined data-types, arrays, ...) can be added as needed. This is achieved by splitting the graphical representation of the new data-type into a set of waveforms of already supported types. We implement this mapping by defining an additional `sc_trace()` function that takes a constant reference to an object of the new type as an argument. In example 10.3.4 we have an aggregate data-type consisting of three fields. The `sc_trace()` function is overloaded such that we get separate waveforms for the fields where each one has the name of the original object plus a postfix consisting of a dot and the name of the field. Providing this single method allows for tracing variables, signals, and ports of a user-defined data-type. It should be noted that in order to transport a user-defined data-type `T` over a `sc_signal` the following two operators *must* be defined

```
bool T::operator==(const T&)
ostream& operator<<(ostream&, const T&)
```

These operators are not required for tracing, though. Hence, as long as no signals are involved, we get by without implementing these operators. Also, a user-defined assignment operator is required if the one provided by the C++ compiler (memberwise copy; see [33]) is inadequate.

Example 10.3.4 *Tracing user-defined aggregate data-types*

```
struct data {
    unsigned char raw;
```

```
    bool parity;
    bool valid;
    bool operator==(const data& other) {
        return other.raw==raw && other.parity==parity &&
            other.valid==valid;
    }
};
ostream& operator<<(ostream& o, const data& d) {
    return o << d.raw << '/' << d.parity << '/' << d.valid;
}
...
void sc_trace(sc_trace_file* tf, const data& d,
              const sc_string& name)
{
    sc_trace(tf, d.raw,    name + ".raw");
    sc_trace(tf, d.parity, name + ".parity");
    sc_trace(tf, d.valid,  name + ".valid");
}
...
data my_data;
sc_signal<data> my_data_sig;
sc_trace(trace_file, my_data, "my_data");
sc_trace(trace_file, my_data_sig, "my_data_sig");
...
```

Support for tracing vectors can be implemented in a similar way. Example 10.3.5 depicts an implementation that is based on a function template. As shown, this approach works even for arrays of signals of user-defined data-types.

Example 10.3.5 *Support for tracing vectors*

```
template <class T>
void my_vector_trace(sc_trace_file* tf, const T x[],
                     const sc_string& name, int length)
{
    for (int i=0; i<length; i++)
        sc_trace(tf, x[i], name + "." +
                           sc_string::to_string("%d", i));
}
...
data data_array[4];
my_vector_trace(trace_file, data_array, "data_array", 4);
...
sc_signal<data> data_sig_array[4];
my_vector_trace(trace_file, data_sig_array, "data_sig_array", 4);
```

The tracing of enumerated types is supported by the SystemC reference implementation. It is not the same for all file formats, though. The code shown in example 10.3.6 runs independently of the file format. However, VCD trace files do not make use of the string representation while WIF trace files do. The enum literals are passed as a null-terminated array of (null-terminated) text strings to `sc_trace()`. Care must be taken not to forget to terminate the array with a `NULL`.

Example 10.3.6 *Tracing enumerated types*

```
enum status {
    idle=0, busy=1, sleeping=2, suspended=3
};
const char* status_as_string[] = { "idle", "busy", "sleeping",
                                   "suspended", NULL };

...
sc_trace_file* trace_file = sc_create_wif_trace_file("trace");
sc_signal<status> status_sig("status_sig");
status  status_var = idle;
sc_trace(trace_file, status_sig, "status_sig", status_as_string);
sc_trace(trace_file, status_var, "status_var", status_as_string);
...
```

A number of additional methods are defined for all trace files. Their behavior depends on the chosen file format.

```
void sc_write_comment(sc_trace_file* tf,
                      const sc_string& comment);
void sc_trace_delta_cycles(sc_trace_file* tf,
                           bool on = true);
```

The first routine allows you to write comments into a trace file. The second one can be used to switch delta-cycle tracing on. Then, *all* value changes are logged in the trace file whereas by default signal value changes are only taken into account when the simulation time is advanced.

Implementing support for additional file formats that allow for storing signal waveforms over time is relatively straightforward. Similar to what has been done for VCD, WIF, and ISDB file formats, we have to create a new trace file class that is derived from `sc_trace_file`, and implement the predefined set of (pure virtual) `trace()` methods. On top of that, we also need methods for creating and closing the new type of trace file (cf. example 10.3.2).

So far we have been concentrating on variables and signals. What about other types of channels? In general, one can pursue a similar approach as before, that is, map the information to a set of waveforms. But not all information fits into the waveform-over-time paradigm. Examples include FIFO buffers, where we might want to see a sequence of samples rather than a waveform[1]. Channels such as `sc_fifo` are also challenging because there can be multiple accesses per delta-cycle. Thus, the tracing mechanism needs to be tightly integrated in order to keep track of the flow of information. We can achieve this by deriving a new class from `sc_fifo`, which has built-in support for file output. The basic idea (see example 10.3.7) is to redefine both `write()` and `nb_write()` so that data is stored in a log file after it has been successfully written into the FIFO.

Example 10.3.7 *Derived FIFO channel with tracing support*

```
#include "systemc.h"
#include "fstream.h"
// Derived FIFO class with added tracing support.
// Writes data into a logfile (separated by newlines) if
// the constructor is called with a non-NULL filename.
template <class T> class my_fifo_t : public sc_fifo<T>
{
public:
    my_fifo_t(const char* nm, int size,
              const char* filename = NULL)
        : sc_fifo<T>(nm, size), log(NULL)
    {
        if (filename != NULL) {
            log = new ofstream(filename);
            if (log->bad())
                cerr << name()
                     << ": (ERROR) could not open file "
                     << filename << endl;
        }
    }

    ~my_fifo_t() { if (log) log->close(); }

    virtual void write(const T& val) {
        sc_fifo<T>::write(val);
        if (log != NULL)
            *log << val << endl;
    }
```

[1]The important difference is that in the waveform case the x-axis denotes the time, which is the same for all channels in the system. If we look at sequences of values, though, then the x-axis gives the sequence index, which may be different for all FIFO channel instances at a given point in time.

```
    virtual bool nb_write(const T& val) {
        bool r = sc_fifo<T>::nb_write(val);
        if (r == true && log != NULL)
            *log << val << endl;
        return r;
    }
protected:
    ofstream* log;
private:
    // Ensure that no one calls the default constructor or
    // the copy constructor by making them private and not
    // implementing them.
    my_fifo_t();
    my_fifo_t(const my_fifo_t&);
};
```

10.4 Debugging

The final section of this chapter is devoted to source code debugging. We present a number of hints that you may find useful in a question-and-answer style. As before, we do not assume the availability of any commercial software. Instead, we focus on using the GNU debugger `gdb`. (These hints usually work regardless of the source code debugger being used, although the commands to be issued may vary from debugger to debugger.) Also, we stick to *documented* interfaces only, and refrain from heavily implementation-dependent "magic" tricks that require changing the source code of the SystemC reference implementation.

Question: How do I get information on the current simulation time while debugging my source code?

Answer: You can make function calls from within a source code debugger. Hence, if you want to know the current simulation time you can call `sc_simulation_time()` when the simulation is stopped in the debugger.

```
Breakpoint 2, decode::entry (this=0xffbe9390) at decode.cpp:69
(gdb) print sc_simulation_time()
$1 = 2
```

If you want the simulation time to be displayed whenever the simulation is stopped then you should use `display` instead of `print`.

```
Breakpoint 2, decode::entry (this=0xffbe9390) at decode.cpp:69
```

```
(gdb) display sc_simulation_time()
1: sc_simulation_time(void) () = 3
(gdb) cont
Continuing.

Breakpoint 2, decode::entry (this=0xffbe9390) at decode.cpp:69
1: sc_simulation_time(void) () = 4
(gdb)
```

Question: How do I make my debugger stop in a particular SystemC process?

Answer: In general, you need to set a breakpoint in the corresponding method.

```
(gdb) break consumer::main
Breakpoint 1 at 0xb0d94: file main.cpp, line 129.
(gdb) run
Starting program: /u/johndoe/systemc/simple_fifo/run.x

             SystemC 2.0 --- Nov  2 2001 17:06:14
        Copyright (c) 1996-2001 by all Contributors
                    ALL RIGHTS RESERVED

Breakpoint 1, consumer::main (this=0x110ee8) at main.cpp:129
(gdb)
```

For method processes (SC_METHOD) this is absolutely sufficient as the method is called once per process activation. Thread processes (SC_THREAD, SC_CTHREAD) are called only once during their first activation, though. A thread process can be suspended by calling wait(). To intercept a resuming thread process breakpoints have to be placed in the source code after all wait() statements. It should be noted that wait() statements can be hidden inside other functions—for instance in blocking access functions such as sc_fifo<T>::read().

Question: Sometimes breakpoints placed on thread processes (SC_THREAD or SC_CTHREAD) don't seem to work. The programs stops only once in the debugger although the process gets activated more often.

Answer: Assume process was a thread process declared inside module. Specifying a breakpoint via

```
(gdb) break module::process
Breakpoint 2 at 0x3f2bc: file module.cpp, line 20.
```

tells the debugger to stop in the first line of executable code inside `module::process`. This line, however, may only be executed once.

```
void module::process() {
    x++; // executed once
    while (1) {
        output() = x++; // executed repeatedly
        wait();
    }
}
```

In the example of the code shown above a breakpoint should be placed on the first line within the `while` statement. Multiple breakpoints are required if `wait()` is called more than once.

```
void module::process() {
    while (1) {
        output() = x++; // set a breakpoint here
        wait();
        output() = y++; // and here too
        wait();
    }
}
```

Question: I have multiple instances of the same module in my design. How do I figure out which one I am in right now?

Answer: If you are in a method of a SystemC module, for instance, in any type of SystemC process, you can call `this->name()` in order to get the instance name of the module.

```
Breakpoint 2, decode::entry (this=0xffbe9390) at decode.cpp:69
1: sc_simulation_time(void) () = 4
(gdb) print this->name()
$2 = 0x1556a1 "DECODE_BLOCK"
(gdb)
```

Question: I have multiple instances of the same module in my design, and I only want the debugger to stop in a particular instance.

Answer: Use a conditional breakpoint in gdb that only stops when code for a particular instance is executed. The condition is that the instance name matches a given text string. For example,

```
(gdb) break numgen::generate if ! strcmp(this->name(), "numgen2")
Breakpoint 2 at 0x3f2bc: file numgen.cpp, line 45.
(gdb) cont
Continuing.
Breakpoint 2, numgen::generate (this=0xffbee7b8) at numgen.cpp:45
45          a -= 1.5;
(gdb) print name()
$1 = 0x1167f9 "numgen2"
(gdb) cont
Continuing.
Breakpoint 2, numgen::generate (this=0xffbee7b8) at numgen.cpp:45
45          a -= 1.5;
(gdb) print name()
$2 = 0x1167f9 "numgen2"
```

Question: I get strange output when I try to make my debugger tell me the value of variables that have SystemC data-types. What should I do?

Answer: C++ source code debuggers are optimized for displaying values of built-in data-types such as `int`, `double`, or `char*`. They do not know about the meaning of a bit-vector or a fixed-point data-type. Hence, when you make them print an instance of such a data-type, they will show you the details of the C++ implementation of this type rather than giving you the information you would like to see: a nicely formatted value. For example:

```
(gdb) print pntr
$1 = {<sc_uint_base> = {num = 0, width = 3}, <No data fields>}
(gdb)
```

The solution is to translate the value of, for instance, a bit-vector into something the debugger understands by calling a conversion function.

```
(gdb) print pntr.to_int()
$2 = 0
(gdb)
```

Examples:

- `to_int()`, `to_uint()`: convert to (unsigned) integer (`sc_bv`, `sc_lv`, integer data-types, fixed-point types)

- to_double(): convert into floating point number (fixed-point data-types, integer types except for sc_int and sc_uint)

- to_string().c_str(): convert into a string representation in a 2-stage process: first, to_string() converts into an sc_string, then c_str() converts this into a const char* that the debugger can display (sc_bv, sc_lv, integer types except for sc_int and sc_uint, fixed-point data-types)

Question: My debugger claims it does not know some conversion functions. What is wrong?

Answer: The C++ compiler may optimize inlined or unused functions away. Make sure functions you want to call from the debugger are not inlined by the compiler by setting the appropriate compiler flags. (For GNU's gcc, the flags are -fno-default-inline -fno-inline.) Also, if the SystemC simulation program does not call these functions they may not be linked in. You may have to force linking them. A simple trick is to have a dummy routine that contains calls to all those conversion functions that you may need. Add to the source code of this routine as your demand grows.

Example 10.4.1 *Force linking of symbols needed for debugging*

```
void dummy_to_force_linking_my_debug_symbols()
{
    // make sure we've got sc_logic::to_char() linked in
    sc_logic l;
    char c = l.to_char();
}
. . .
int sc_main(int argc, char* argv[]) {
    dummy_to_force_linking_my_debug_symbols();
    . . .
}
```

Explicit template instantiation (see [33]) is a more elegant way to achieve the same effect for class and function templates. Putting

```
template class sc_bv<8>;
```

into sc_main() will result in *all* methods of sc_bv<8> being linked into the simulation executable.

Question: Even though I made sure all required functions get linked in I still don't get calling `my_bitvec.to_string().c_str()` to work in the debugger.

Answer: Some debuggers cannot call methods of temporary class objects (like calling the `c_str()` method of the temporary `sc_string` object returned by `to_string()`). A powerful debugging trick is the inclusion of extra functions that are there just to be called from the debugger.

```
template <int wordlength> const char*
bv2s(sc_bv<wordlength>* bitvec) {
    if (bitvec == NULL)
        return "<NULL>";
    static sc_string s = bitvec->to_string();
    return s.c_str();
}

// explicit instantiation of the bv2s() function template
// for 8 and 16 bit
template const char* bv2s(sc_bv<8>*);
template const char* bv2s(sc_bv<16>*);
```

The code above shows a function that converts (a pointer to) a bit-vector into a text string that can be displayed in the debugger. As `bv2s()` is a function template, we need to explicitly instantiate it (unless it is called in the source code) in order to ensure that it is compiled and linked into the simulation executable.

Question: I get strange output when I try to make my debugger tell me the value of a `sc_in<T>` or `sc_out<T>` port. What should I do?

Answer: Again, the C++ source code debugger does not understand the meaning of a port. If you make it print one it will show you the details of a port's C++ implementation.

```
Breakpoint 3, decode::entry (this=0xffbe9390) at decode.cpp:45
(gdb) print destreg_write
$7 = {<sc_port<sc_signal_in_if<bool>,1>> = {<sc_port_b<sc_signal\
_in_if<bool> >> = {<sc_port_base> = {<sc_object> = {static kind_\
string = 0x15df98 "sc_object", m_simc = 0x1c9338, m_name = 0x1c9\
a79 "DECODE_BLOCK.port_5", m_attr_cltn = {m_cltn = {<sc_pvector_\
base> = {m_alloc = 10, m_sz = 0, m_data = 0x1dc518}, <No data fi\
elds>}}, _vptr. = 0x1c6e90}, static kind_string = 0x160a80 "sc_p\
ort_base", m_bind_info = 0x0}, m_interface = 0xffbeb430, m_inter\
face_vec = {<sc_pvector_base> = {m_alloc = 10, m_sz = 1, m_data \
```

```
= 0x1dc618}, <No data fields>}}, <No data fields>}, static kind_\
string = 0x161618 "sc_in", m_traces = 0x0}
```

Use the port's `read()` method to get the value of the connected signal instead. (Note that this approach does not work for `sc_fifo` as `read()` changes its state. Normally this is OK since you don't need to debug unread items within an `sc_fifo` channel, but if you really need to do this you can create a specialized version of `sc_fifo` that has a method that allows items within the FIFO to be read without being consumed.)

```
(gdb) print destreg_write.read()
$8 = (bool &) @0xffbeb45c: false
(gdb)
```

Question: What if my signal has a SystemC data-type?

Answer: Combine the approaches described above. You can first use the `read()` method to get the value and then call a conversion function to translate it into something the source code debugger can print well.

Question: What about `printf()`?

Answer: `printf()` (or any other type of console or file output) can be a great debugging tool—especially when it comes to keeping track of parallel processes. Without jeopardizing the synthesizability of a design one can easily embed some instrumentation using preprocessor directives such as `#ifdef DEBUG`. When generating debug traces make sure to include simulation time, instance name, and location in the file. A macro like the one shown in the next example may come in handy.

Example 10.4.2 *Instrumenting code with debug output*

```
#define DEBUG_TRACE                                          \
cout << "[t=" << sc_simulation_time()                        \
     << ",l=" << __LINE__ << "] ";                           \
{                                                            \
    sc_simcontext* simc = sc_get_curr_simcontext();          \
    if(simc!=0 && simc->is_running()) {                      \
        sc_process_b* p = simc->get_curr_proc_info()         \
                                ->process_handle;            \
        if(p != 0) cout << p->name();                        \
    }                                                        \
}                                                            \
cout << ": "
```

```
SC_MODULE(my_module) {
    . . .
    void my_process() {
        . . .
        #ifdef DEBUG
        DEBUG_TRACE << "the answer is " << 42 << endl;
        #endif
        . . .
    }
};
```

Question: My SystemC simulation aborts with an error message. How can I find out where exactly the problem is located?

Answer: Before issuing a warning or an error message the SystemC simulator calls `sc_stop_here()`. The sole purpose of this function is to serve as a place to put a breakpoint. Hence, start the debugger, set a breakpoint on `sc_stop_here()`, and run the simulation. Once the breakpoint is hit maneuver upward in the call stack (e.g., by calling `up` in `gdb`) until you see where the problem is.

10.5 Summary

In this chapter we have discussed a variety of practical techniques for constructing testbenches in SystemC, tracing data during simulation, and debugging SystemC designs. It is likely that most of these techniques will prove valuable in almost all SystemC designs since a great deal of effort is typically required in the debugging and verification process, regardless of the modeling levels or design methodology used in a particular case.

11

Conclusions and the Future of SystemC

11.1 Summary of the Book

So, what have we learned about SystemC in this book?

We have covered many aspects of SystemC, with reference to the current (2.0) version of the language and reference implementation. These included:

- Modeling abstraction levels, including functional and platform-level models
- Fundamental constructs of SystemC, including time, modules, interfaces, ports, channels (both primitive and hierarchical), method and thread processes, events, sensitivity, instantiation, and simulation semantics
- Some discussion of models of computation including Kahn process networks and dataflow, and how to model them in SystemC
- Classical hardware modeling in SystemC, at the RT level, with a comprehensive example, and at the behavioral level, along with a summary of hardware-oriented data-types
- Parameterized modules and channels, including parameter resolution issues and techniques for protecting intellectual property
- Interface and channel design techniques

- Transaction-level modeling, including techniques for writing very high speed models
- Communication refinement, including a discussion of the various options for how to do this, and examples of hardware–hardware, software–software, and hardware–software refinement
- Testbenches, tracing and debugging, with practical guidelines on debugging questions

The best way for you to become comfortable applying the modeling concepts introduced in this book is to experiment with some SystemC models and designs, focusing on modeling issues or refinement steps that are important in your particular case. In this sort of new endeavor, sometimes making the first step is the most difficult, but one of SystemC's great strengths is that it is easy to get started with it, assuming you have some knowledge of C++. (And if you don't have much knowledge of C++, there are many good C++ books and courses available.) There are excellent, free C++ compilers and debuggers available (e.g., see *gcc.gnu.org*) that run on a wide range of platforms, including PCs; there is the free SystemC reference simulator (see *www.systemc.org*); the examples from this book can be downloaded for free (see instructions in chapter 1); and example designs and an active discussion forum are available at *www.systemc.org*.

In addition, many EDA vendors are now offering sophisticated tool support for SystemC, including co-simulation capabilities with Verilog and VHDL, and RTL and behavioral synthesis tools that work directly with SystemC input. To see some of the tools that are available for SystemC, visit the *Products and Solutions* page at *www.systemc.org*.

So, what are you waiting for? With SystemC, the world awaits you!

11.2 The Future of SystemC

OSCI has a number of initiatives underway to advance SystemC in the future. As of the writing of this book (March 2002), it has formed several working groups [35] to evolve the language and supporting methodologies:

1. The Language Working Group, to be the central focus for the core language evolution, standardization, the reference implementation of the simulator, and decision making on language additions

2. A Verification Working Group, looking at library extensions on top of the core language to better support the development of intelligent testbenches, transaction-based verification methods, and advanced verification platforms

3. A Dataflow Working Group, proposed to add library extensions on top of the core language to support dataflow modeling (going beyond what is outlined in chapter 3 and chapter 5)

4. An IP Integration Working Group, looking at methods and guidelines for modeling IP blocks within SystemC and packaging them for effective reuse

5. An Analog-Mixed-Signal Working Group, looking at extensions to SystemC to support modeling of analog blocks and mixed-signal systems integrating both digital and analog components

6. A Transaction API Working Group, looking at defining interfaces between SystemC models and simulations and other simulation and hardware emulation environments, employing a transaction-level abstraction and associated API for the interface

In addition, the SystemC roadmap, with the Language Working Group providing oversight and guidance, includes the following possible areas for consideration in extending the SystemC core language:

- dynamic thread creation, for possible use in modeling software tasks and real-time operating system behavior; fork and join, for better modeling of concurrency; interrupt and abort modeling support; performance modeling support; and modeling support for timing specification and constraints

- enhanced support for embedded software modeling, including abstract RTOS modeling and scheduler modeling

- analog and mixed-signal extensions and their integration with the SystemC core language

Clearly, the development of SystemC in the future will depend heavily on inputs from the associated working groups and heavy feedback and input from users on requirements for future evolution.

11.3 In Conclusion

We hope that you have found this book valuable and that it helps you in the process of learning and applying SystemC to your system-level design problems. We hope that this book and other books, papers and presentations by SystemC experts will help the SystemC community of users grow in numbers and in their understanding.

Bibliography

Each entry ends with a list of the pages on which it is cited.

[1] B. Bailey, G. de Jong, P. Schaumont, and C. Lennard. Interface based design: Using the VSI system-level interface behavioral documentation standard. In *Forum on Design Languages*, pages 221–230, Tübingen, 2000. ⟨3, 4⟩

[2] F. Balarin, M. Chiodo, P. Giusto, H. Hsieh, A. Jurecska, L. Lavagno, C. Passerone, A. Sangiovanni-Vincentelli, E. Sentovich, K. Suzuki, and B. Tabbara. *Hardware–Software Co-Design of Embedded Systems: The Polis Approach.* Kluwer Academic Publishers, 1997. ⟨3⟩

[3] S. S. Bhattacharyya, P. K. Murthy, and E. A. Lee. *Software Synthesis from Dataflow Graphs.* Kluwer Academic Publishers, 1996. ⟨47⟩

[4] I. Bolsens, H. De Man, B. Lin, K. Van Rompaey, S. Vercauteren, and D. Verkest. Hardware/software co-design of digital telecommunication systems. *Proceedings of the IEEE*, 85(3):391–418, 1997. ⟨159⟩

[5] J. T. Buck. *Scheduling Dynamic Dataflow Graphs with Bounded Memory.* PhD thesis, University of California at Berkeley, 1993. ⟨78, 173⟩

[6] D. R. Butenhof. *Programming with POSIX Threads.* Addison-Wesley, Reading, MA, 1997. ⟨168⟩

[7] H. Chang, L. Cooke, M. Hunt, A. McNelly, G. Martin, and L. Todd. *Surviving the SOC Revolution: A Guide to Platform-Based Design.* Kluwer Academic Publishers, 1999. ⟨1⟩

[8] E. A. de Kock, G. Essink, W. J. M. Smits, P. van der Wolf, J.-Y. Brunel, W. M. Kruijtzer, P. Lieverse, and K. A. Vissers. YAPI: Application modeling for signal processing systems. In *Proceedings of the Design Automation Conference*, pages 402–405, 2000. ⟨46⟩

[9] S. Devadas, A. Ghosh, and K. Keutzer. *Logic Synthesis.* McGraw-Hill, New York, NY, 1994. ⟨51⟩

[10] S. Edwards, L. Lavagno, E. A. Lee, and A. Sangiovanni-Vincentelli. Design of embedded systems: Formal models, validation and synthesis. *Proceedings of the IEEE*, 85(3):366–390, March 1997. ⟨41⟩

[11] European SystemC Users Group. Web page at URL: *http://www-ti.informatik.uni-tuebingen.de/~systemc.* ⟨5⟩

[12] D. D. Gajski, J. Zhu, R. Dömer, A. Gerstlauer, and S. Zhao. *SpecC: Specification Language and Methodology.* Kluwer Academic Publishers, 2000. ⟨3, 4⟩

[13] ISO/IEC 14882 C++ Standard, 1998. ⟨61⟩

[14] G. Kahn. The semantics of a simple language for parallel programming. In J. L. Rosenfeld, editor, *Information Processing*. North-Holland Publishing Company, 1974. ⟨46, 78⟩

[15] G. Kahn. Coroutines and networks of parallel processes. In *Information Processing*. North-Holland Publishing Company, 1977. ⟨78⟩

[16] D. E. Knuth. *The Art of Computer Programming*, volume 1. Addison-Wesley, Reading, MA, 3rd edition, 1997. ⟨26⟩

[17] E. A. Lee. Consistency in dataflow graphs. In *IEEE Transactions on Parallel and Distributed Systems*, April 1991. ⟨78, 173⟩

[18] E. A. Lee and D. G. Messerschmitt. Static scheduling of synchronous dataflow programs for digital signal processing. In *IEEE Transactions on Computers*, January 1987. ⟨78, 173⟩

[19] E. A. Lee and D. G. Messerschmitt. Synchronous dataflow. *Proceedings of the IEEE*, 75:1235–1245, 1987. ⟨47⟩

[20] E. A. Lee and T. M. Parks. Dataflow process networks. *Proceedings of the IEEE*, 83:773–801, 1987. ⟨78⟩

[21] C. K. Lennard, P. Schaumont, G. de Jong, A. Haverinen, and P. Hardee. Standards for system-level design: practical reality or solution in search of a question? In *Proceedings of the Design Automation & Test Conference in Europe*, pages 576–583, Paris, 2000. ⟨3, 4⟩

[22] S. Liao, S. Tjiang, and R. Gupta. An efficient implementation of reactivity for modeling hardware in the Scenic design environment. In *Proceedings of the Design Automation Conference*, pages 70–75, Anaheim, 1997. ⟨4⟩

[23] R. Lipsett, C. F. Schaefer, and C. Ussery. *VHDL: Hardware Description and Design*. Kluwer Academic Publishers, 1993. ⟨3, 37⟩

[24] S. Meyers. *Effective C++: 50 Specific Ways to Improve Your Program and Design*. Addison-Wesley, Reading, MA, 2nd edition, 1997. ⟨3⟩

[25] G. De Micheli. *Synthesis and Optimization of Digital Circuits*. McGraw-Hill, New York, NY, 1994. ⟨49⟩

[26] G. De Micheli, R. Ernst, and W. Wolf, editors. *Readings in Hardware/Software Co-Design*. Morgan Kaufmann, San Francisco, CA, 2001. ⟨3⟩

[27] P. R. Moorby and D. E. Thomas. *The Verilog Hardware Description Language*. Kluwer Academic Publishers, 1998. ⟨3⟩

[28] D. Ramanathan, R. Roth, and R. Gupta. Interfacing hardware and software using C++ class libraries. In *Proceedings of the International Conference on Computer Design*, pages 445–450, 2000. ⟨5⟩

[29] K. Van Rompaey, D. Verkest, I. Bolsens, and H. De Man. CoWare—a design environment for heterogeneous hardware/software systems. In *Proceedings of the European Design Automation Conference*, pages 252–257, 1996. ⟨4⟩

[30] J. A. Rowson and A. Sangiovanni-Vincentelli. Interface-based design. In *Proceedings of the Design Automation Conference*, pages 178–183, 1997. ⟨23⟩

[31] J. Rumbaugh, M. Blaha, W. Premerlani, F. Eddy, and W. Lorensen. *Object-Oriented Modeling and Design.* Prentice Hall, Englewood Cliffs, NJ, 1991. ⟨17⟩

[32] P. Schaumont, S. Vernalde, L. Rijnders, M. Engels, and I. Bolsens. Synthesis of multi-rate and variable rate circuits for high speed telecommunications applications. In *Proceedings of the European Design and Test Conference*, pages 542–546, 1997. ⟨5⟩

[33] B. Stroustrup. *The C++ Programming Language.* Addison-Wesley, Reading, MA, special edition, 2000. ⟨3, 18, 103, 191, 199⟩

[34] SystemC User's Forum Presentation, June 7, 2000. Available from the Open SystemC Initiative (OSCI) at *http://www.systemc.org/news/systemc_user_forum.pdf.* ⟨4⟩

[35] SystemC User's Forum Presentation, June 17, 2001. Available from the Open SystemC Initiative (OSCI) at *http://www.systemc.org/news/systemc_dac_01.pdf.* ⟨4, 204⟩

[36] *SystemC Version 2.0 Beta-1 User's Guide*, 2001. Available from the Open SystemC Initiative (OSCI) at *http://www.systemc.org.* ⟨3, 70⟩

[37] *Functional Specification for SystemC 2.0*, January 2001. Available from the Open SystemC Initiative (OSCI) at *http://www.systemc.org.* ⟨3⟩

[38] Virtual Socket Interface Alliance (VSIA). URL: *http://www.vsi.org.* ⟨23⟩

```
    my_module abc("abc");
    ...
    sc_trace(tf, abc.input,     "abc.input");
    sc_trace(tf, abc.output,    "abc.output");
    sc_trace(tf, abc.inout,     "abc.inout");
    sc_trace(tf, abc.clk,       "abc.clk");
    sc_trace(tf, abc.local_sig, "abc.local_sig");
    sc_trace(tf, abc.state,     "abc.state");
    // For signed/unsigned char/short/int/long we can also
    // specify a bit-width. Here, we trace everything again
    // as 2-bit.
    sc_trace(tf, abc.input,     "abc.input2",     2);
    sc_trace(tf, abc.output,    "abc.output2",    2);
    sc_trace(tf, abc.inout,     "abc.inout2",     2);
    sc_trace(tf, abc.local_sig, "abc.local_sig2", 2);
    sc_trace(tf, abc.state,     "abc.state2",     2);
    ...
}
```

Support for tracing additional data-types (user-defined data-types, arrays, ...) can be added as needed. This is achieved by splitting the graphical representation of the new data-type into a set of waveforms of already supported types. We implement this mapping by defining an additional `sc_trace()` function that takes a constant reference to an object of the new type as an argument. In example 10.3.4 we have an aggregate data-type consisting of three fields. The `sc_trace()` function is overloaded such that we get separate waveforms for the fields where each one has the name of the original object plus a postfix consisting of a dot and the name of the field. Providing this single method allows for tracing variables, signals, and ports of a user-defined data-type. It should be noted that in order to transport a user-defined data-type `T` over a `sc_signal` the following two operators *must* be defined

```
bool T::operator==(const T&)
ostream& operator<<(ostream&, const T&)
```

These operators are not required for tracing, though. Hence, as long as no signals are involved, we get by without implementing these operators. Also, a user-defined assignment operator is required if the one provided by the C++ compiler (memberwise copy; see [33]) is inadequate.

Example 10.3.4 *Tracing user-defined aggregate data-types*

```
struct data {
    unsigned char raw;
```

This book was set by the authors in Robert Slimbach's Minion and Charles Bigelow and Chris Holmes' `Lucida Sans Typewriter` using $\LaTeX 2_\varepsilon$, with a slightly modified version of Stephen A. Edwards's book template.

About the Authors

Thorsten Grötker is an engineering manager in the System-Level Design Automation Group of Synopsys, Inc. Prior to joining Synopsys in 1997, he worked in a research position for the Integrated Systems for Signal Processing Lab. He holds an electrical engineering and a doctorate degree, both from the Aachen University of Technology, Aachen, Germany.

Stan Liao is a Principal Engineer in the Advanced Technology Group of Synopsys, Inc., where he has been working since 1996. He holds a Ph.D. in Electrical Engineering and Computer Science from Massachusetts Institute of Technology.

Grant Martin is a Fellow in the Labs of Cadence Design Systems. He has also worked for Burroughs Machines Limited in Scotland and for Bell-Northern Research/Northern Telecom in Canada. He has a Bachelor's and Master's degree in Mathematics (Combinatorics and Optimization) from the University of Waterloo, Canada.

Stuart Swan is a Senior Architect in the System Design and Verification group at Cadence Design Systems, where he has been working since 1991. Previously he worked for Valid Logic Systems and Zilog, Inc. He has a Bachelor's degree in Electrical Engineering from Stanford University.

About the Authors